Selected Titles in This Series

Volume

12 **I. G. Macdonald**
Symmetric functions and orthogonal polynomials
1998

11 **Lars Gårding**
Some points of analysis and their history
1997

10 **Victor Kac**
Vertex algebras for beginners
1997

9 **Stephen Gelbart**
Lectures on the Arthur-Selberg trace formula
1996

8 **Bernd Sturmfels**
Gröbner bases and convex polytopes
1996

7 **Andy R. Magid**
Lectures on differential Galois theory
1994

6 **Dusa McDuff and Dietmar Salamon**
J-holomorphic curves and quantum cohomology
1994

5 **V. I. Arnold**
Topological invariants of plane curves and caustics
1994

4 **David M. Goldschmidt**
Group characters, symmetric functions, and the Hecke algebra
1993

3 **A. N. Varchenko and P. I. Etingof**
Why the boundary of a round drop becomes a curve of order four
1992

2 **Fritz John**
Nonlinear wave equations, formation of singularities
1990

1 **Michael H. Freedman and Feng Luo**
Selected applications of geometry to low-dimensional topology
1989

Symmetric Functions and Orthogonal Polynomials

University
LECTURE
Series

Volume 12

Symmetric Functions and Orthogonal Polynomials

Dean Jacqueline B. Lewis Memorial Lectures
Rutgers University

I. G. Macdonald

American Mathematical Society
Providence, Rhode Island

1991 *Mathematics Subject Classification.* Primary 05, 22, 33.

Library of Congress Cataloging-in-Publication Data

Macdonald, I. G. (Ian Grant)
Symmetric functions and orthogonal polynomials / I. G. Macdonald.
p. cm. — (University lecture series, ISSN 1047-3998 ; v. 12)
"Dean Jacqueline B. Lewis Memorial lectures, Rutgers University."
Includes bibliographical references.
ISBN 0-8218-0770-6 (softcover : alk. paper)
1. Symmetric functions. 2. Orthogonal polynomials. I. Title. II. Series: University lecture series (Providence, R. I.) ; 12.
QA212.M33 1997
515′.22–dc21 97-26100
CIP

∞ The paper used in this book is acid-free and falls within the guidelines
established to ensure permanence and durability.
Visit the AMS home page at URL: `http://www.ams.org/`

10 9 8 7 6 5 4 3 2 1 02 01 00 99 98

Contents

Preface

The first two chapters of these notes reproduce, with additional detail here and there, the content of the oral lectures delivered in March 1993. For example, I have included the original proof of the existence theorem (Chapter II, §§6 and 7) which previously was available only in preprint form.

The postscript (Chapter III) perhaps requires a word of explanation. During the period that has elapsed since these lectures were given, the subject has enjoyed considerable advance and clarification, mainly thanks to the work of Ivan Cherednik, who perceived that the affine Hecke algebras provided a key to unlock its mysteries. Chapter III provides a brief survey of these recent developments.

Introduction

The aim of this Introduction is to provide some background and perspective to the subject of these lectures for the benefit of readers who are not already well-versed in these matters and to explain its connections with other branches of mathematics, such as combinatorics and representation theory. References in square brackets are to the bibliography at the end of this Introduction, *not* to that at the end of the book.

Symmetric functions

The theory of symmetric functions is one of the most classical parts of algebra, going back to the 16th and 17th centuries and the attempts of mathematicians of that epoch to solve polynomial equations of degree higher than the second. The coefficients of a polynomial equation in one unknown are, up to sign, just the elementary symmetric functions of the roots of that equation, and hence any symmetric polynomial in the roots can be expressed uniquely as a polynomial in the coefficients. For example, the equations

$$ne_n = \sum_{r=1}^{n} (-1)^{r-1} p_r e_{n-r},$$

which serve to express the power sums p_r in terms of the elementary symmetric functions e_r, were first found by Isaac Newton in the 17th century. The first part of Chapter I (§§1–6) provides a rapid survey, from a modern viewpoint, of this classical material.

Schur functions and their generalizations

Of the various families of symmetric functions, the most significant are undoubtedly the Schur functions, because of their intimate relationship with the irreducible characters of both the symmetric groups and the general linear groups, and for their combinatorial applications.

Let us consider the representation-theoretic aspect first.

(i) The irreducible (complex) characters χ^λ of the symmetric group S_n are indexed by the partitions λ of n. So also are the conjugacy classes in S_n, because the conjugacy class of a permutation is determined by its cycle-type, which is a partition of n. The value χ^λ_μ of the character χ^λ at an element of cycle-type μ is then given by the scalar product

$$\chi^\lambda_\mu = \langle s_\lambda, p_\mu \rangle.$$

Equivalently, χ^λ_μ is the coefficient of s_λ in the expression of p_μ as a sum of Schur functions.

(ii) A matrix representation ρ of the general linear group $\mathrm{GL}_n(\mathbb{C})$ is said to be *polynomial* if the matrix coefficients $\rho_{ij}(X)$, $X \in \mathrm{GL}_n(\mathbb{C})$, are polynomial functions of the entries of the matrix X. The irreducible polynomial representations ρ_λ of $\mathrm{GL}_n(\mathbb{C})$ are indexed by the partitions $\lambda = (\lambda_1, \ldots, \lambda_n)$ of length $\leq n$, and the character of ρ_λ is

$$\operatorname{trace} \rho_\lambda(X) = s_\lambda(\xi_1, \ldots, \xi_n)$$

where $\xi_1, \ldots, \xi_n$ are the eigenvalues of the matrix $X \in \mathrm{GL}_n(\mathbb{C})$.

The Schur functions also arise naturally in various combinatorial contexts: not only those associated with the combinatorics of the symmetric groups, but also, for example, in the enumeration of plane partitions. A *plane partition* may be thought of as an infinite matrix $\pi = (\pi_{ij})_{i,j\geq 1}$ in which each entry π_{ij} is a nonnegative integer, such that $\pi_{ij} \geq \pi_{i+1,j}$ and $\pi_{ij} \geq \pi_{i,j+1}$ for all (i,j), and such that $|\pi| = \sum_{i,j} \pi_{ij}$ is finite (so that all but a finite number of the π_{ij} are zero). Thus a plane partition is not so very different from a tableau, as defined in Chapter I, §7; and the expression of the Schur function s_λ as a sum of monomials indexed by tableaux (loc. cit.) can therefore be exploited, as was first realized by R. Stanley [**S**], to derive generating functions for various classes of plane partitions, such as (for example) MacMahon's famous generating function for all plane partitions:

$$\sum_\pi t^{|\pi|} = \prod_{n\geq 1} (1-t^n)^{-n}.$$

The Schur functions lend themselves readily to generalization in various directions, and some (but by no means all) of these generalizations are briefly surveyed in §§8–12 of Chapter I. Of these, both the zonal polynomials (§8) and the Hall-Littlewood functions (§10) have close connections with representation theory: they may be interpreted as the values of zonal spherical functions on the general linear group $\mathrm{GL}_n(F)$, where F is a local field, relative to a maximal compact subgroup K. In the case of the zonal polynomials, F is the real field and K is the orthogonal group; and in the case of the Hall-Littlewood functions, F is a p-adic field and $K = \mathrm{GL}_n(\mathfrak{o})$, where $\mathfrak{o}$ is the ring of integers in F (and the parameter t is equal to $1/q$, where q is the cardinality of the residue field of F). Moreover, just as the Schur functions determine the character tables of the symmetric groups, as we have seen above, so the Hall-Littlewood functions play a crucial role in the determination of the irreducible complex characters of the finite general linear groups, as was first realized by J. A. Green [**G1**].

Finally, both the zonal polynomials and the Hall-Littlewood functions are subsumed as limiting cases of the two-parameter symmetric functions that are the subject of §§11 and 12. As remarked at the end of §12, the structure of some of the formulas suggests strongly that these symmetric functions should be regarded as attached to root systems of type A and that they should have counterparts attached to other root systems.

This more general theory is the subject of Chapter II. Apart from symmetric functions, there are two other forerunners of this theory, namely the theory of hypergeometric functions and Jacobi polynomials attached to root systems developed by G. Heckman and E. Opdam [**HO**], and the succession of constant term conjectures (now, fortunately, all theorems) starting with those of F. J. Dyson [**D**] and G. E. Andrews [**A**]. We consider these inputs in turn.

Jacobi polynomials attached to root systems

The Jacobi polynomials $P_\lambda(k)$ of Heckman and Opdam (loc. cit.) attached to a root system R are a limiting case of our $P_\lambda(q,t)$, where the parameters q and t (thought of as real or complex numbers) both tend to 1 in such a way that $(1-t)/(1-q)$ tends to a definite limit k. They were defined in loc. cit. as the simultaneous eigenfunctions of a second-order differential operator that extrapolates the radial part of the Laplace operator on a symmetric space G/K, where G is a connected noncompact semisimple Lie group and K is a maximal compact subgroup of G; the root system R is the restricted root system of G/K, and the parameter k is half the root multiplicity (assumed constant, for simplicity of description). When the root system is of type A, the Jacobi polynomials coincide with the Jack symmetric functions of Chapter I, §9.

Constant term identities

The constant term identities now to be described constitute another antecedent of the orthogonal polynomials of Chapter II, of a more combinatorial flavour. In 1962 F. J. Dyson [**D**] was led by considerations of statistical mechanics to conjecture that the constant term in the expansion of the product

$$\prod_{\substack{i,j=1 \\ i\neq j}}^{n} (1-x_i x_j^{-1})^k,$$

where $x_1,\dots,x_n$ are independent variables and k is a positive integer, should be

$$(nk)!/(k!)^n. \tag{1}$$

This conjecture was soon proved true, by J. Gunson [**G2**] and K. Wilson [**W**], who showed more generally that the constant term in the expansion of

$$\prod_{i\neq j}(1-x_i x_j^{-1})^{k_i}, \tag{2}$$

where $k_1,\dots,k_n$ are nonnegative integers, is the multinomial coefficient

$$\frac{(k_1+\cdots+k_n)!}{k_1!\cdots k_n!}. \tag{3}$$

Next, in 1975 G. E. Andrews [**A**] conjectured a q-analogue of this result. Write

$$(x;q)_n = \prod_{i=1}^{n}(1-q^{i-1}x),$$

where q,x are independent variables and n is an integer ≥ 0. Then Andrews' conjecture was that the polynomial ($\in \mathbb{Z}[x_1^{\pm1},\dots,x_n^{\pm1},q]$)

$$\prod_{i\neq j}(\varepsilon_{ij}x_i x_j^{-1};q)_{k_i}, \tag{4}$$

where $\varepsilon_{ij}=1$ or q according as $i<j$ or $i>j$, should have constant term (i.e., independent of $x_1,\dots,x_n$) equal to

$$\frac{(q;q)_{k_1+\cdots+k_n}}{(q;q)_{k_1}\cdots(q;q)_{k_n}} \tag{5}$$

when $q=1$, this reduces to the previous result. However, (5) proved a harder nut to crack, and was finally proved correct by Zeilberger and Bressoud [**ZB**] in 1985.

These results and conjectures in turn prompted analogous conjectures [**M**] for an arbitrary root system R. To state these in their simplest form, assume that R is reduced and irreducible, and for each $\alpha \in R$ let e^α denote the corresponding formal exponential, as in Chapter II, §3. Then the constant term in the expansion of the product

$$\prod_{\alpha \in R} (1 - e^\alpha)^k \tag{6}$$

(where as before k is a positive integer) should be

$$\prod_{i=1}^{r} \binom{kd_i}{k}, \tag{7}$$

where $d_1, \dots, d_r$ are the degrees of the fundamental polynomial invariants of the Weyl group of R; and, more generally, the constant term (i.e., involving q but no exponentials) in

$$\prod_{\alpha \in R^+} (e^\alpha; q)_k (qe^{-\alpha}; q)_k \tag{8}$$

should be

$$\prod_{i=1}^{r} \begin{bmatrix} kd_i \\ k \end{bmatrix}, \tag{9}$$

where $\left[\begin{smallmatrix} n \\ r \end{smallmatrix} \right]$ is the q-binomial coefficient or Gaussian polynomial, defined by

$$\begin{bmatrix} n \\ r \end{bmatrix} = (q;q)_n / (q;q)_r (q;q)_{n-r}.$$

Clearly (9) implies (7) by setting $q = 1$. Moreover, when the root system R is of type A_{n-1}, the degrees d_i are $2, 3, \dots, n$, and so (9) reduces to the special case of (5) in which $k_1 = \dots = k_n = k$. Whether the general form of (5), with n distinct parameters k_i, has counterparts for other root systems is still an open question.

The combinatorialists made some progress towards establishing (9) on a case-by-case basis—see [**K**] for the classical root systems, [**GG**] for F_4 and [**H**] for G_2—but were unable to handle the cases of the exceptional root systems E_6, E_7, and E_8. The first uniform proof, for all root systems R, of the conjecture (7) (the case $q = 1$ of (9)) was found by E. Opdam [**O**], by exploiting the shift operators he and Heckman had constructed as part of their theory of hypergeometric functions and Jacobi polynomials alluded to earlier in this introduction; and finally (9) was proved in full generality by Cherednik [**C**] in 1995.

In the context of the theory of orthogonal polynomials attached to root systems developed in Chapter II, the conjecture (9) essentially appears as the simplest case ($\lambda = 0$) of the norm formula (8.3′). (The function Δ (with $t = q^k$) defined in Chapter II, §4, is not quite the same as (8) above, but it is not difficult to switch from one to the other.)

Finally, it should be said that the account in Chapter II of these notes reflects the state of knowledge at the time (March 1993) the lectures were delivered. At that time, for example, the norm formula referred to above was still conjectural. In Chapter III I have attempted to remedy this defect to some extent, by surveying some of the more recent developments involving the affine Hecke algebra, such as the nonsymmetric orthogonal polynomials E_λ (Chapter III, §6), a proof of the norm formula (and hence in particular of (9) above) in §7, and Cherednik's Fourier transform (§9). However, as the subject is still evolving rapidly, this survey is necessarily incomplete.

References

[A] G. E. Andrews, *Problems and prospects for basic hypergeometric functions*, in Theory and Applications of Special Functions, edited by R. Askey, Academic Press, New York, 1975, pp. 191–224.

[C] I. Cherednik, *Double affine Hecke algebras and Macdonald's conjectures*, Ann. Math. **141** (1995), 191–216.

[D] F. J. Dyson, *Statistical theory of the energy levels of complex systems*. I, J. Math. Phys. **3** (1962), 140–156.

[GG] F. Garvan and G. Gonnet, *Macdonald's constant term conjectures for exceptional root systems*, Bull. Amer. Math. Soc. (N.S.) **24** (1991), 343–347.

[G1] J. A. Green, *The characters of the finite general linear groups*, Trans. Amer. Math. Soc. **80** (1955), 402–447.

[G2] J. Gunson, *Proof of a conjecture of Dyson in the statistical theory of energy levels*, J. Math. Phys. **3** (1962), 752–753.

[H] L. Habsieger, *La q-conjecture de Macdonald–Morris pour G_2*, C. R. Acad. Sci. Paris Sér. I Math. **303** (1986), 211–213.

[HO] G. Heckman and E. Opdam, *Root systems and hypergeometric functions*. I–IV, Comp. Math. **64** (1987), 329–352, 353–373; ibid. **67** (1988), 21–49, 191–209.

[K] K. Kadell, *A proof of the q-Macdonald-Morris conjecture for BC_n*, Mem. Amer. Math. Soc. **108** (1994), no. 516.

[M] I. G. Macdonald, *Some conjectures for root systems*, SIAM J. Math. Anal. **13** (1982), 988–1007.

[O] E. Opdam, *Some applications of hypergeometric shift operators*, Inv. Math. **98** (1989), 1–18.

[S] R. P. Stanley, *Theory and application of plane partitions*. I, II, Studies in Appl. Math. **50** (1971), 167–188, 259–279.

[W] K. Wilson, *Proof of a conjecture by Dyson*, J. Math. Phys. **3** (1962), 1040–1043.

[ZB] D. Zeilberger and D. Bressoud, *A proof of Andrews' q-Dyson conjecture*, Discrete Math. **54** (1985), 201–224.

CHAPTER 1

Symmetric Functions

1. The ring of symmetric functions

Let $x_1, \dots, x_n$ be independent indeterminates. The symmetric group S_n acts on the polynomial ring $\mathbb{Z}[x_1, \dots, x_n]$ by permuting the x's , and we shall write

$$\Lambda_n = \mathbb{Z}[x_1, \dots, x_n]^{S_n}$$

for the subring of symmetric polynomials in $x_1, \dots, x_n$. If $f \in \Lambda_n$, we may write

$$f = \sum_{r \geq 0} f^{(r)},$$

where $f^{(r)}$ is the homogeneous component of f of degree r. Each $f^{(r)}$ is itself symmetric, and so Λ_n is a *graded* ring:

$$\Lambda_n = \bigoplus_{r \geq 0} \Lambda_n^r,$$

where Λ_n^r is the additive group of homogeneous symmetric polynomials of degree r in $x_1, \dots, x_n$. (By convention, 0 is homogeneous of every degree.)

If we now adjoin another indeterminate x_{n+1}, we can form $\Lambda_{n+1} = \mathbb{Z}[x_1, \dots, x_{n+1}]^{S_{n+1}}$, and we have a surjective homomorphism (of graded rings)

$$\Lambda_{n+1} \to \Lambda_n, \tag{1.1}$$

defined by setting $x_{n+1} = 0$. The mapping $\Lambda_{n+1}^r \to \Lambda_n^r$ is surjective for all $r \geq 0$, and bijective if and only if $r \leq n$.

Often it is convenient to pass to the limit. Let

$$\Lambda^r = \varprojlim_n \Lambda_n^r$$

for each $r \geq 0$, and let

$$\Lambda = \bigoplus_{r \geq 0} \Lambda^r.$$

By the definition of inverse (or projective) limits, an element of Λ^r is a sequence $(f_n)_{n \geq 0}$ where $f_n \in \Lambda_n^r$ for each n, and f_n is obtained from f_{n+1} by setting $x_{n+1} = 0$. We may therefore regard the f_n as the partial sums of an infinite series f of monomials of degree r in infinitely many indeterminates $x_1, x_2, \dots$. For example, if $f_n = x_1 + \cdots + x_n$ for each n, then $f = \sum_{i=1}^{\infty} x_i$. Thus the elements of Λ are no longer polynomials, and are traditionally called symmetric *functions*.

For each n there is a surjective homomorphism $\Lambda \to \Lambda_n$, obtained by setting $x_m = 0$ for all $m > n$. The graded ring Λ is the *ring of symmetric functions*. If R is any commutative ring, we shall write

$$\Lambda_R = \Lambda \otimes_{\mathbb{Z}} R, \quad \Lambda_{n,R} = \Lambda_n \otimes_{\mathbb{Z}} R$$

for the ring of symmetric functions (resp. symmetric polynomials in n indeterminates) with coefficients in R.

In the following sections of this chapter we shall survey various bases of the ring of symmetric functions. They will all be indexed by *partitions*. A partition is a (finite or infinite) sequence

$$\lambda = (\lambda_1, \lambda_2, \lambda_3, \dots)$$

of integers, such that $\lambda_1 \geq \lambda_2 \geq \cdots \geq 0$, and

$$|\lambda| = \sum \lambda_i < \infty$$

so that from a certain point onwards (if the sequence λ is infinite) all the λ_i are zero. We shall not distinguish between two such sequences which differ only by a string of zeros at the end. Thus $(2,1)$, $(2,1,0)$, and $(2,1,0,0,\dots)$ are all to be regarded as the same partition.

The nonzero λ_i are called the *parts* of λ, and the number of parts is the *length* $l(\lambda)$ of λ. If λ has m_1 parts equal to 1, m_2 parts equal to 2, and so on, we shall occasionally write $\lambda = (1^{m_1}2^{m_2}\cdots)$, although strictly speaking we should write this in reverse order.

Let $\mathcal{P}$ denote the set of all partitions, and $\mathcal{P}_n$ the set of all partitions of n (i.e., such that $|\lambda| = n$). The *natural* (or *dominance*) partial order on $\mathcal{P}$ is defined as follows.

(1.2) $\lambda \geq \mu$ *if and only if* $|\lambda| = |\mu|$ *and* $\lambda_1 + \cdots + \lambda_r \geq \mu_1 + \cdots + \mu_r$ *for all* $r \geq 1$.

It is a total order on $\mathcal{P}_n$ for $n \leq 5$, but not for $n \geq 6$.

With each partition λ we associate a *diagram*, consisting of the points $(i,j) \in \mathbb{Z}^2$ such that $1 \leq j \leq \lambda_i$. We adopt the convention (as with matrices) that the first coordinate i (the row index) increases as one goes downwards, and the second coordinate j (the column index) increases from left to right. Often it is more convenient to replace the lattice points (i,j) by squares, and then the diagram of λ consists of λ_1 boxes in the top row, λ_2 boxes in the second row, and so on; the whole arrangement of boxes being left-justified.

If we read the diagram of a partition λ by columns, we obtain the *conjugate* partition λ'. Thus λ'_j is the number of boxes in the jth column of the diagram of λ, and hence is equal to the number of parts of λ that are $\geq j$. Equivalently, the diagram of λ' is obtained from that of λ by reflection in the main diagonal. It is not difficult to show that

$$\lambda \geq \mu \Leftrightarrow \mu' \geq \lambda'.$$

2. Monomial symmetric functions

Let $\lambda = (\lambda_1, \lambda_2, \dots)$ be a partition. It determines a monomial

$$x^\lambda = x_1^{\lambda_1} x_2^{\lambda_2} \cdots.$$

The *monomial symmetric function* m_λ is the sum of all distinct monomials that can be obtained from x^λ by permutations of the x's. For example, $m_{(21)} = \sum x_i^2 x_j$, summed over all pairs (i,j) such that $i \neq j$.

For each partition λ, the image $m_\lambda(x_1, \dots, x_n)$ of m_λ in Λ_n (obtained by setting $x_m = 0$ for all $m > n$) is zero if and only if $l(\lambda) > n$. As λ runs through the partitions of length $\leq n$, the $m_\lambda(x_1, \dots, x_n)$ form a $\mathbb{Z}$-basis of Λ_n, and as λ runs through all partitions the m_λ form a $\mathbb{Z}$-basis of Λ.

3. Elementary symmetric functions

When $\lambda = (1^r)$, m_λ is the rth *elementary symmetric function* e_r:

$$e_r = m_{(1^r)} = \sum_{i_1 < \cdots < i_r} x_{i_1} \cdots x_{i_r}.$$

(For $r = 0$, we define e_0 to be 1.) The e_r have the generating function

$$E(t) = \sum_{r \geq 0} e_r t^r = \prod_{i \geq 1} (1 + x_i t) \tag{3.1}$$

where t is another indeterminate.

For each partition $\lambda = (\lambda_1, \lambda_2, \ldots)$ let

$$e_\lambda = e_{\lambda_1} e_{\lambda_2} \cdots .$$

Then the e_λ form another $\mathbb{Z}$-basis of Λ as λ runs through all partitions. Equivalently, we have

(3.2) $\Lambda = \mathbb{Z}[e_1, e_2, \ldots]$, *and* $e_1, e_2, \ldots$ *are algebraically independent over* $\mathbb{Z}$.

PROOF. Consider the product $e_{\lambda'}$, where λ' is the conjugate of λ. When multiplied out it is a sum of monomials of the form

$$(x_{i_1} x_{i_2} \cdots)(x_{j_1} x_{j_2} \cdots) \cdots = x^\alpha,$$

say, where $i_1 < i_2 < \cdots < i_{\lambda'_1}$, $j_1 < j_2 < \cdots < j_{\lambda'_2}$, and so on. If we now enter the numbers $i_1, i_2, \ldots, i_{\lambda'_1}$ in order down the first column of the diagram of λ, then the numbers $j_1, j_2, \ldots, j_{\lambda'_2}$ in order down the second column, and so on, it is clear that for each $r \geq 1$ all the numbers $\leq r$ so entered in the diagram of λ must occur in the top r rows. Hence $\alpha_1 + \cdots + \alpha_r \leq \lambda_1 + \cdots + \lambda_r$ for each $r \geq 1$, and it follows that $e_{\lambda'}$ is of the form

$$e_{\lambda'} = m_\lambda + \sum_{\mu < \lambda} a_{\lambda\mu} m_\mu \tag{3.3}$$

with nonnegative integer coefficients $a_{\lambda\mu}$. The triangular form of these equations shows that they can be solved to give m_λ as a linear combination of the e_μ. □

4. Complete symmetric functions

For each $r \geq 1$, the rth *complete symmetric function* h_r is

$$h_r = \sum_{|\lambda| = r} m_\lambda,$$

the sum of all monomials of degree r. (For $r = 0$, we define h_0 to be 1.) The generating function of the h_r is

$$H(t) = \sum_{r \geq 0} h_r t^r = \prod_{i \geq 1} (1 - x_i t)^{-1}, \tag{4.1}$$

as one sees by expanding each factor $(1 - x_i t)^{-1}$ as a geometric series and then multiplying these series together. From (3.1) and (4.1) it follows that

$$H(t)E(-t) = 1,$$

i.e., that

$$\sum_{r=0}^{n}(-1)^r e_r h_{n-r} = 0 \tag{4.2}$$

for each $n \geq 1$. Since the e_r are algebraically independent we may define a ring homomorphism $\omega\colon \Lambda \to \Lambda$ by

$$\omega(e_r) = h_r \quad (r \geq 1)$$

and the symmetry of the relations (4.2) shows that $\omega(h_r) = e_r$ for each r, so that $\omega^2 = 1$, i.e., ω is an involution. From (3.2) it follows that

(4.3) $\Lambda = \mathbb{Z}[h_1, h_2, \ldots]$, *and* $h_1, h_2, \ldots$ *are algebraically independent over* $\mathbb{Z}$.

Equivalently, the products

$$h_\lambda = h_{\lambda_1} h_{\lambda_2} \cdots = \omega(e_\lambda)$$

form a third $\mathbb{Z}$-basis of Λ.

5. Power sums

For each $r \geq 1$, the rth *power sum* is

$$p_r = m_{(r)} = \sum x_i^r.$$

Their generating function is

$$P(t) = \sum_{r\geq 1} p_r t^{r-1} = \sum_i \frac{x_i}{1 - x_i t}, \tag{5.1}$$

i.e.,

$$P(t) = H'(t)/H(t) \tag{5.2}$$

by (4.1), where $H'(t)$ is the derivative of $H(t)$ with respect to t. If we write this in the form

$$H'(t) = P(t)H(t)$$

and pick out the coefficient of t^{n-1} on either side, we shall obtain

$$nh_n = \sum_{r=1}^{n} p_r h_{n-r} \tag{5.3}$$

for each $n \geq 1$. It follows from (5.3), by induction on n, that $h_n \in \mathbb{Q}[p_1, \ldots, p_n]$ and $p_n \in \mathbb{Z}[h_1, \ldots, h_n]$. Hence

(5.4) $\Lambda_{\mathbb{Q}} = \mathbb{Q}[p_1, p_2, \ldots]$ *and* $p_1, p_2, \ldots$ *are algebraically independent over* $\mathbb{Q}$.

Equivalently, the power-sum products

$$p_\lambda = p_{\lambda_1} p_{\lambda_2} \cdots$$

form a $\mathbb{Q}$-basis of $\Lambda_{\mathbb{Q}}$. However, they do *not* form a $\mathbb{Z}$-basis of Λ: for example, $e_2 = \frac{1}{2}(p_1^2 - p_2)$ does not have integral coefficients as a linear combination of the p_λ.

Likewise, from (5.1) and (3.1) we have

$$P(-t) = E'(t)/E(t). \tag{5.5}$$

Since the involution ω interchanges $E(t)$ and $H(t)$, it follows from (5.2) and (5.5) that ω interchanges $P(t)$ and $P(-t)$, i.e., we have

$$\omega(p_r) = (-1)^{r-1} p_r \tag{5.6}$$

for all $r \geq 1$.

Finally, we may compute h_n as a polynomial in the power sums, as follows: by (5.2) we have

$$P(t) = \frac{d}{dt} \log H(t)$$

and therefore

$$\begin{aligned} H(t) &= \exp\left(\sum_{r\geq 1} \frac{p_r t^r}{r} \right) \\ &= \prod_{r\geq 1} \exp \frac{p_r t^r}{r} \\ &= \prod_{r\geq 1} \left(\sum_{m_r \geq 0} \frac{1}{m_r!} \left(\frac{p_r t^r}{r} \right)^{m_r} \right). \end{aligned}$$

Let us pick out the coefficient of p_λ in this product. If $\lambda = (1^{m_1} 2^{m_2} \cdots)$, the coefficient is z_λ^{-1}, where

$$z_\lambda = \prod_{r\geq 1} (r^{m_r} \cdot m_r!) \tag{5.7}$$

and we have

$$h_n = \sum_{|\lambda|=n} z_\lambda^{-1} p_\lambda. \tag{5.8}$$

This numerical function z_λ (which will occur frequently in this chapter) has the following interpretation. Let $|\lambda| = n$ and let $w \in S_n$ be a permutation of cycle-type λ (i.e., the product of disjoint cycles of lengths $\lambda_1, \lambda_2, \ldots$). Then z_λ is the order of the centralizer of w in S_n.

6. Scalar product

Let $x = (x_1, x_2, \ldots)$ and $y = (y_1, y_2, \ldots)$ be two (finite or infinite) sequences of independent indeterminates. For each $f \in \Lambda$ we shall denote the corresponding symmetric function of the x_i (resp. the y_j, resp. the products $x_i y_j$) by $f(x)$ (resp. $f(y)$, resp. $f(xy)$). Thus for example

$$p_n(xy) = \sum_{i,j} (x_i y_j)^n = p_n(x) p_n(y)$$

and therefore

$$p_\lambda(xy) = p_\lambda(x) p_\lambda(y) \tag{6.1}$$

for all partitions λ.

Consider now the product

$$\Pi(x,y) = \prod_{i,j} (1 - x_i y_j)^{-1}.$$

We have

$$\begin{aligned}\Pi(x,y) &= \sum_{n\geq 0} h_n(xy) \qquad \text{by (4.1)}\\ &= \sum_\lambda z_\lambda^{-1} p_\lambda(xy) \quad \text{by (5.8)}\end{aligned}$$

and therefore by (6.1)

(6.2) $$\Pi(x,y) = \sum_\lambda z_\lambda^{-1} p_\lambda(x) p_\lambda(y).$$

Another expansion of $\Pi(x,y)$ may be obtained as follows: we have

$$\begin{aligned}\Pi(x,y) &= \prod_j H(y_j)\\ &= \prod_j \sum_{\alpha_j\geq 0} h_{\alpha_j}(x) y_j^{\alpha_j}\end{aligned}$$

from which it follows that

(6.3) $$\begin{aligned}\Pi(x,y) &= \sum_\lambda h_\lambda(x) m_\lambda(y)\\ &= \sum_\lambda m_\lambda(x) h_\lambda(y)\end{aligned}$$

by interchanging the x's and the y's.

We shall now define a scalar product on Λ, i.e., a $\mathbb{Z}$-valued bilinear form $\langle f,g\rangle$, by requiring that the bases (h_λ) and (m_λ) be dual to each other, i.e., that

(6.4) $$\langle h_\lambda, m_\mu\rangle = \delta_{\lambda\mu}$$

for all partitions λ and μ, where $\delta_{\lambda\mu}$ is the Kronecker delta.

(6.5) *For each $n\geq 0$ let $(u_\lambda), (v_\lambda)$ be $\mathbb{Q}$-bases of $\Lambda^n_{\mathbb{Q}}$, indexed by the partitions of n. Then the following conditions are equivalent*:

(a) $\langle u_\lambda, v_\mu\rangle = \delta_{\lambda\mu}$ *for all* λ,μ,
(b) $\sum_\lambda u_\lambda(x) v_\lambda(y) = \Pi(x,y)$.

PROOF. Let

$$u_\lambda = \sum_\rho a_{\lambda\rho} h_\rho, \quad v_\mu = \sum_\sigma b_{\mu\sigma} m_\sigma.$$

Then

$$\langle u_\lambda, v_\mu\rangle = \sum_\rho a_{\lambda\rho} b_{\mu\rho}$$

so that (a) is equivalent to

(a′) $$\sum_\rho a_{\lambda\rho} b_{\mu\rho} = \delta_{\lambda\mu}.$$

Also (b) is equivalent to the identity

$$\sum_\lambda u_\lambda(x) v_\lambda(y) = \sum_\rho h_\rho(x) m_\rho(y)$$

by (6.3), hence to

(b′) $$\sum_\lambda a_{\lambda\rho} b_{\lambda\sigma} = \delta_{\rho\sigma}.$$

Since (a′) and (b′) are equivalent, so are (a) and (b). □

From (6.5) and (6.2) it follows that

(6.6) $$\langle p_\lambda, p_\mu \rangle = \delta_{\lambda\mu} z_\lambda,$$

which shows that the scalar product defined by (6.4) is symmetric and positive definite, and that the p_λ form an orthogonal basis of $\Lambda_{\mathbb{Q}}$. Finally, it follows from (5.6) that $\omega p_\lambda = \pm p_\lambda$ for each partition λ, and hence from (6.6) that

(6.7) *The involution ω is an isometry, i.e. $\langle \omega u, \omega v \rangle = \langle u, v \rangle$ for all $u, v \in \Lambda$.*

7. Schur functions

For the moment we shall work in the polynomial ring $\mathbb{Z}[x_1, \ldots, x_n]$. For each sequence $\alpha = (\alpha_1, \ldots, \alpha_n)$ of nonnegative integers, let $x^\alpha = x_1^{\alpha_1} \cdots x_n^{\alpha_n}$ and let

$$\begin{aligned} a_\alpha &= \det(x_i^{\alpha_j})_{1 \le i,j \le n} \\ &= \sum_{w \in S_n} \varepsilon(w) w(x^\alpha) \end{aligned}$$

where $\varepsilon(w) = \pm 1$ is the sign of the permutation w. Thus if $\delta = (n-1, n-2, \ldots, 1, 0)$, a_δ is the Vandermonde determinant, equal to the product $\prod_{i<j}(x_i - x_j)$.

Clearly a_α will merely change sign if two of the α_i are interchanged, and in particular will vanish if two are equal. Up to sign, we may therefore assume that $\alpha_1 > \alpha_2 > \cdots > \alpha_n \ge 0$, and write $\alpha_i = \lambda_i + n - i$ for $1 \le i \le n$, that is to say $\alpha = \lambda + \delta$ where $\lambda = (\lambda_1, \ldots, \lambda_n)$ is a *partition* (of length at most n). The determinant $a_\alpha = a_{\lambda+\delta}$ is divisible by the Vandermonde determinant a_δ in the polynomial ring $\mathbb{Z}[x_1, \ldots, x_n]$, and the quotient

(7.1) $$s_\lambda(x_1, \ldots, x_n) = a_{\lambda+\delta}/a_\delta$$

is a *symmetric* polynomial, homogeneous of degree $|\lambda|$. It is easily seen that on passing to $n+1$ variables we have $s_\lambda(x_1, \ldots, x_n, 0) = s_\lambda(x_1, \ldots, x_n)$, and it follows that there is a uniquely defined element $s_\lambda \in \Lambda$ that reduces to $s_\lambda(x_1, \ldots, x_n)$ when $x_{n+1}, x_{n+2}, \ldots$ are all set equal to zero, for any $n \ge l(\lambda)$. This symmetric function s_λ is the *Schur function* corresponding to λ.

We may remark that the definition (7.1) makes sense for any sequence $\lambda = (\lambda_1, \ldots, \lambda_n)$ of nonnegative integers, not necessarily a partition. Since $a_{\lambda+\delta} = \varepsilon(w) a_{w(\lambda+\delta)}$ for any $w \in S_n$ it follows that $s_\lambda = 0$ if the sequence $\lambda + \delta$ has two or more equal terms, and that otherwise $s_\lambda = \varepsilon(w) s_\mu$ where μ is a partition and $w \in S_n$ is uniquely determined by $\mu + \delta = w(\lambda + \delta)$.

(7.2) *Let λ be a partition. Then*

$$s_\lambda = m_\lambda + \sum_{\mu < \lambda} K_{\lambda\mu} m_\mu$$

for suitable coefficients $K_{\lambda\mu}$.

PROOF. We shall again work with n variables $x_1, \dots, x_n$, where $n \geq |\lambda|$. Consider the product $m_\lambda a_\delta$: we have

$$m_\lambda a_\delta = \sum_{\mu,w} \varepsilon(w) x^{\mu+w\delta}$$

summed over $w \in S_n$ and permutations μ of λ. In this sum we may replace μ by $w\mu$, so that

$$m_\lambda a_\delta = \sum_\mu a_{\mu+\delta}$$

and therefore

$$m_\lambda = \sum_\mu s_\mu$$

in Λ_n for all large n, and therefore also in Λ. Now it is easily checked that if μ is a permutation of λ and $\mu \neq \lambda$, then s_μ is either 0 or is equal to $\pm s_\nu$ where $\nu < \lambda$. Hence m_λ is of the form

$$m_\lambda = s_\lambda + \sum_{\nu<\lambda} a_{\lambda\nu} s_\nu .$$

On inverting these equations we obtain (7.2). □

REMARK. In fact the coefficients $K_{\lambda\mu}$ (which are called Kostka numbers) are positive integers and have the following combinatorial interpretation (see e.g. [**M6**] Chapter I, §6). If λ is any partition, a *tableau of shape* λ is any mapping τ of the diagram of λ into the positive integers such that $\tau(i,j)$ increases in the weak sense along the rows of the diagram and in the strict sense down the columns. The *weight* of τ is the sequence $(\tau^{-1}(r))_{r\geq 1}$, which may or may not be a partition. With this explained, we have

(7.3) *$K_{\lambda\mu}$ is the number of tableaux of shape λ and weight μ.*

From (7.2) it follows in particular that the s_λ form another $\mathbb{Z}$-basis of Λ. In fact they form an *orthonormal* basis relative to the scalar product (6.4), i.e., we have

(7.4) $$\langle s_\lambda, s_\mu \rangle = \delta_{\lambda\mu} \text{ for all partitions } \lambda, \mu.$$

Equivalently, by (6.5):

(7.5) $$\sum_\lambda s_\lambda(x) s_\lambda(y) = \Pi(x,y).$$

PROOF. There are many ways of proving (7.5). One way uses Cauchy's determinant: if $x = (x_1, \dots, x_n)$ and $y = (y_1, \dots, y_n)$ then we have

(7.6) $$\det\left(\frac{1}{x_i + y_j}\right)_{1\leq i,j\leq n} = \frac{a_\delta(x) a_\delta(y)}{\prod_{i,j}(x_i + y_j)}.$$

To prove (7.6) we observe that the determinant on the left, multiplied by the product of all the $x_i + y_j$, is a homogeneous polynomial $C(x,y)$ of total degree $n^2 - n$. On the other hand, $C(x,y)$ is skew-symmetric in the x's and in the y's, hence is divisible in $\mathbb{Z}[x,y]$ by $a_\delta(x)a_\delta(y)$, which is also homogeneous of total degree $n^2 - n$. It follows that $C(x,y) = c a_\delta(x) a_\delta(y)$ for some integer c, which may be shown to be 1 by comparing the coefficients of $x^\delta y^\delta$ on either side.

If we now replace each x_i by $-x_i^{-1}$ in (7.6) we obtain an equivalent version:

$$\det\left(\frac{1}{1-x_iy_j}\right)_{1\le i,j\le n} = a_\delta(x)a_\delta(y)\Pi(x,y). \tag{7.7}$$

On the other hand we have

$$\begin{aligned}\det\left(\frac{1}{1-x_iy_j}\right) &= \det\left(\sum_{\alpha_j\ge 0} x_i^{\alpha_j}y_j^{\alpha_j}\right)\\ &= \sum_\alpha a_\alpha(x)y^\alpha\end{aligned}$$

summed over all sequences $\alpha=(\alpha_1,\dots,\alpha_n)$ of nonnegative integers. Hence

$$\det\left(\frac{1}{1-x_iy_j}\right) = \sum_\lambda a_{\lambda+\delta}(x)a_{\lambda+\delta}(y) \tag{7.8}$$

summed over all partitions of length $\le n$. The identity (7.5) now follows from (7.7) and (7.8), by letting $n\to\infty$. □

We may remark that the Schur functions s_λ are uniquely determined by (7.2) and (7.4). Indeed, they are overdetermined by these two conditions. For the Gram-Schmidt process shows that they are uniquely determined by (7.2) and the requirement (weaker than (7.4)) that $\langle s_\lambda, s_\mu\rangle = 0$ if $\lambda > \mu$.

The Schur functions s_λ can of course be expressed in terms of the other bases of Λ that we have encountered. First we have

(7.9) *If λ is a partition of length $\le n$ then*

$$s_\lambda = \det(h_{\lambda_i-i+j})_{1\le i,j\le n}$$

with the convention that $h_r=0$ if $r<0$ (*and* $h_0=1$).

PROOF. Again, there are many proofs of this formula. We shall deduce it from (7.5). If $x=(x_1,\dots,x_n)$ and $y=(y_1,\dots,y_n)$ as before, let $H(t)=\sum_{r\ge 0}h_r(x)t^r$, so that

$$\Pi(x,y) = \prod_{i=1}^n H(y_i)$$

and therefore

$$\begin{aligned}\Pi(x,y)a_\delta(y) &= \det(y_i^{n-j}H(y_i))\\ &= \det\left(\sum_{\alpha_j\ge 0} h_{\alpha_j}(x)y_i^{\alpha_j+n-j}\right)\\ &= \sum_\alpha h_\alpha(x)a_{\alpha+\delta}(y)\end{aligned}$$

summed over all sequences $\alpha=(\alpha_1,\dots,\alpha_n)$ of nonnegative integers. Hence for each partition λ of length $\le n$, the coefficient of $a_{\lambda+\delta}(y)$ in $\Pi(x,y)a_\delta(y)$ is

$$\sum_{w\in s_n}\varepsilon(w)h_{w(\lambda+\delta)-\delta} = \det(h_{\lambda_i-i+j})_{1\le i,j\le n}.$$

On the other hand, it follows from (7.5) that the coefficient of $a_{\lambda+\delta}(y)$ in $\Pi(x,y)a_\delta(y)$ is $s_\lambda(x)$. □

In terms of the elementary symmetric functions, we have

(7.10) $$s_\lambda = \det(e_{\lambda'_i-i+j})_{1\le i,j\le m}$$

where $\lambda' = (\lambda'_1, \lambda'_2, \dots)$ *is the conjugate of* λ, *and* $m \ge l(\lambda')$. (As with the h's, we define e_r to be 0 when $r < 0$.)

PROOF. Consider the matrices

$$H = (h_{i-j})_{1\le i,j\le N}, \qquad E = ((-1)^{i-j}e_{i-j})_{1\le i,j\le N},$$

where $N = m+n$. Both H and E are triangular, with 1's down the diagonal, hence each has determinant equal to 1. Moreover, the relations (4.2) show that $H = E^{-1}$. Hence each minor of H is equal to the complementary cofactor of E', the transpose of E. Now the determinant in (7.9) is an $n \times n$ minor of H, corresponding to the rows (in increasing order) $\lambda_n+1, \lambda_{n-1}+2, \dots, \lambda_1+n$, and the first n columns. The complementary cofactor of E' turns out to be precisely the determinant (7.10). □

Finally, it follows from (7.9) and (7.10) that

(7.11) $$\omega s_\lambda = s_{\lambda'} \textit{ for all partitions } \lambda.$$

In the remaining sections of this chapter, we shall survey (rather sketchily) various classes of symmetric functions that are analogues or generalizations of the Schur functions.

8. Zonal polynomials ([M6], Chapter VII)

These are certain symmetric functions Z_λ, indexed as usual by partitions, which (when restricted to a finite number of variables $x_1, \dots, x_n$ and partitions λ of length $\le n$) arise naturally in connection with Fourier analysis on the homogeneous space G/K, where $G = \mathrm{GL}_n(\mathbb{R})$ and K is the orthogonal group $O(n)$ (so that G/K may be identified, via $X \mapsto XX^t$, with the space of positive definite real symmetric $n \times n$ matrices). It should be remarked that the Schur functions s_λ arise in exactly the same way when $G = \mathrm{GL}_n(\mathbb{C})$ and K is the unitary group $U(n)$.

The Z_λ, suitably normalized, are characterized by the following two properties, which are the counterparts of (7.2) and (7.4):

(8.1) $$Z_\lambda = m_\lambda + \text{ lower terms},$$

where by "lower terms" is meant a linear combination of the m_μ such that $\mu < \lambda$; and

(8.2) $$\langle Z_\lambda, Z_\mu\rangle_2 = 0 \quad \text{if } \lambda \ne \mu,$$

where the scalar product $\langle u, v\rangle_2$ on Λ is defined by

(8.3) $$\langle p_\lambda, p_\mu\rangle = \delta_{\lambda\mu} \cdot 2^{l(\lambda)} z_\lambda$$

with z_λ as defined in (5.7), and $l(\lambda)$ the length of λ.

As in §7, these two properties (8.1) and (8.2) overdetermine the Z_λ, and their existence therefore requires proof. We shall not give a proof at this stage, since the Z_λ are limiting cases of a more general family of symmetric functions to be considered later (§11). Unlike the Schur functions, there is no closed formula such as (7.1), (7.9), or (7.10) available to give a direct definition of the Z_λ.

9. Jack's symmetric functions ([M6], Chapter VI, §10)

These are symmetric functions $P_\lambda^{(\alpha)}$, indexed by partitions and depending rationally on a parameter α (so that they lie in Λ_F, where F is the field $\mathbb{Q}(\alpha)$), which interpolate between the Schur functions s_λ and the zonal polynomials Z_λ. They are characterized by the following two properties:

$$P_\lambda^{(\alpha)} = m_\lambda + \text{ lower terms,} \tag{9.1}$$

$$\langle P_\lambda^{(\alpha)}, P_\mu^{(\alpha)} \rangle_\alpha = 0 \quad \text{if } \lambda \neq \mu, \tag{9.2}$$

where the scalar product is now defined by

$$\langle p_\lambda, p_\mu \rangle_\alpha = \delta_{\lambda\mu} \cdot \alpha^{l(\lambda)} z_\lambda. \tag{9.3}$$

Clearly when $\alpha = 1$ they reduce to the Schur functions s_λ, and when $\alpha = 2$ to the zonal polynomials Z_λ. They also have well-defined limits as $\alpha \to 0$ and $\alpha \to \infty$ (even though the scalar product (9.3) collapses), and in fact

$$P_\lambda^{(\alpha)} \to e_{\lambda'} \quad \text{as } \alpha \to 0,$$

$$P_\lambda^{(\alpha)} \to m_\lambda \quad \text{as } \alpha \to \infty.$$

Also when $\alpha = \frac{1}{2}$ they occur in nature, as zonal spherical functions on the homogeneous space G/K, where now $G = \mathrm{GL}_n(\mathbb{H})$ and K is the quaternionic unitary group of $n \times n$ matrices ([M6], Chapter VII, §6).

10. Hall-Littlewood symmetric functions ([M6], Chapter III)

These symmetric functions arose originally in connection with the combinatorial and enumerative lattice properties of finite abelian groups. Unlike the zonal polynomials and Jack's symmetric functions, they can be defined by a closed formula. Let $x_1, \ldots, x_n$ and t be indeterminates, and for each partition $\lambda = (\lambda_1, \ldots, \lambda_n)$ of length $\leq n$ define

$$P_\lambda(x_1, \ldots, x_n; t) = v_\lambda(t)^{-1} \sum_{w \in S_n} w \left(x^\lambda \prod_{i<j} \frac{x_i - t x_j}{x_i - x_j} \right), \tag{10.1}$$

where $v_\lambda(t)$ is a normalizing factor designed to ensure that the coefficient of x^λ in P_λ is 1; explicitly

$$v_\lambda(t) = \prod_{i \geq 0} \prod_{j=1}^{m_i} \frac{1 - t^j}{1 - t}, \tag{10.2}$$

where m_i is the number of terms in the sequence $(\lambda_1, \ldots, \lambda_n)$ equal to i. When P_λ is written as a sum of monomials, the coefficients are polynomials in t, i.e., $P_\lambda(x_1, \ldots, x_n; t) \in \Lambda_n[t]$.

It is not difficult to show that on passing to $n + 1$ variables we have $P_\lambda(x_1, \ldots, x_n, 0; t) = P_\lambda(x_1, \ldots, x_n; t)$ if $l(\lambda) \leq n$, and hence that there is a uniquely defined element $P_\lambda(t) \in \Lambda$ that reduces to $P_\lambda(x_1, \ldots, x_n; t)$ when $x_{n+1}, x_{n+2}, \ldots$ are all set equal to zero, for any $n \geq l(\lambda)$. This symmetric function $P_\lambda(t)$ is the *Hall-Littlewood function* corresponding to the partition λ.

The Hall-Littlewood functions serve to interpolate between the Schur functions s_λ and the monomial symmetric functions m_λ: for it follows from the definitions (10.1) and (10.2) that $P_\lambda(0) = s_\lambda$ and $P_\lambda(1) = m_\lambda$ for each partition λ.

The $P_\lambda(t)$ are characterized by the following two properties:

(10.3) $$P_\lambda(t) = m_\lambda + \text{ lower terms},$$

(10.4) $$\langle P_\lambda(t), P_\mu(t)\rangle_{(t)} = 0 \text{ if } \lambda \neq \mu,$$

where the scalar product is now defined by

(10.5) $$\langle p_\lambda, p_\mu\rangle_{(t)} = \delta_{\lambda\mu} z_\lambda \prod_{i=1}^{l(\lambda)} (1 - t^{\lambda_i})^{-1}.$$

11. The symmetric functions $P_\lambda(q,t)$

The symmetric functions surveyed in the last four sections, namely the Schur functions s_λ, the zonal polynomials Z_λ, the Jack symmetric functions $P_\lambda^{(\alpha)}$ and the Hall-Littlewood functions $P_\lambda(t)$, are all special cases of a family of symmetric functions $P_\lambda = P_\lambda(q,t)$, indexed as usual by partitions and depending rationally on two parameters q and t (so that they lie in Λ_F, where $F = \mathbb{Q}(q,t)$ is the field of rational functions in q and t). We may regard q and t either as independent indeterminates or as real variables.

The P_λ are characterized by the same two conditions as before:

(11.1) $$P_\lambda = m_\lambda + \text{ lower terms},$$

(11.2) $$\langle P_\lambda, P_\mu\rangle_{q,t} = 0 \quad \text{if } \lambda \neq \mu,$$

where the scalar product is now defined by

(11.3) $$\langle P_\lambda, P_\mu\rangle_{q,t} = \delta_{\lambda\mu} z_\lambda \prod_{i=1}^{l(\lambda)} \frac{1 - q^{\lambda_i}}{1 - t^{\lambda_i}}.$$

As we have observed in previous sections, these two conditions overdetermine the P_λ, and their existence therefore requires proof. Before we come to the proof, let us consider some particular cases.

(1) When $q = t$, the scalar product (11.3) reduces to that defined in §6, and hence $P_\lambda(q,q)$ is the Schur function s_λ (independent of q).

(2) When $q = 0$, (11.3) reduces to (10.5) and hence $P_\lambda(0,t)$ is the Hall-Littlewood function $P_\lambda(t)$.

(3) Let $q = t^\alpha$ where $\alpha \in \mathbb{R}, \alpha > 0$, and let $t \to 1$, so that $q \to 1$ also. Then

$$\frac{1 - q^m}{1 - t^m} = \frac{1 - t^{\alpha m}}{1 - t^m} \to \alpha$$

as $t \to 1$, for all $m \geq 1$. Hence the scalar product (11.3) tends to (9.3) as $t \to 1$, and consequently

$$\lim_{t \to 1} P_\lambda(t^\alpha, t)$$

is the Jack symmetric function $P_\lambda^{(\alpha)}$.

(4) When $t = 1$ (and q is arbitrary) we have $P_\lambda(q,1) = m_\lambda$.

(5) When $q = 1$ (and t is arbitrary) we have $P_\lambda(1,t) = e_{\lambda'}$.

(6) Finally, it is clear from (11.3) that

$$\langle p_\lambda, p_\mu\rangle_{q^{-1},t^{-1}} = q^{-n} t^n \langle p_\lambda, p_\mu\rangle_{q,t}$$

if λ and μ are partitions of n. Hence the scalar products $\langle u, v\rangle_{q,t}$ and $\langle u, v\rangle_{q^{-1},t^{-1}}$ are proportional on each homogeneous summand Λ_F^n of Λ_F, and therefore $P_\lambda(q^{-1}, t^{-1}) = P_\lambda(q,t)$.

If we regard q and t as real numbers, we may summarize these special cases in the following diagram, in which the point (q,t) represents the basis $(P_\lambda(q,t))$ of Λ_F (or of $\Lambda_{\mathbb{R}}$).

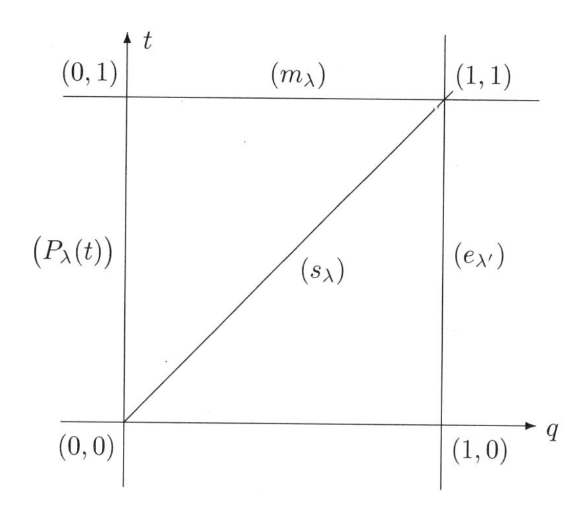

At each point (q,q) on the diagonal of the square we have the Schur functions s_λ, at each point on the upper edge ($t=1$) the monomial symmetric functions m_λ, and so on. In this scheme the Jack symmetric functions, for varying α, correspond to the points in the infinitesimal neighbourhood of the point $(1,1)$, and more precisely $(P_\lambda^{(\alpha)})$ corresponds to the direction through $(1,1)$ with slope $1/\alpha$. Notice that the bottom edge ($t=0$) of the square remains unmarked; I do not know if the $P_\lambda(q,0)$ have any reasonable interpretation (except that they can be derived from the Hall-Littlewood functions $P_\lambda(0,t)$ by duality (see later)).

Let $x=(x_1,x_2,\dots)$ and $y=(y_1,y_2,\dots)$ be two sequences of independent indeterminates and define

$$\Pi = \Pi(x,y;q,t) = \prod_{i,j} \frac{(tx_iy_j;q)_\infty}{(x_iy_j;q)_\infty} \tag{11.4}$$

where

$$(a;q)_\infty = \prod_{r=0}^{\infty}(1-aq^r)$$

for any a for which the product on the right makes sense. Then we have

$$\Pi(x,y;q,t) = \sum_\lambda z_\lambda(q,t)^{-1} p_\lambda(x) p_\lambda(y) \tag{11.5}$$

where

$$z_\lambda(q,t) = z_\lambda \prod_{i=1}^{l(\lambda)} \frac{1-q^{\lambda_i}}{1-t^{\lambda_i}}.$$

PROOF. We have

$$\begin{aligned}\log \Pi &= \sum_{i,j}\sum_{r=0}^{\infty}(\log(1-tx_iy_jq^r)-\log(1-x_iy_jq^r))\\ &= \sum_{i,j}\sum_{r\geq 0}\sum_{n\geq 1}\frac{1}{n}(x_iy_jq^r)^n(1-t^n)\\ &= \sum_{n\geq 1}\frac{1}{n}\frac{1-t^n}{1-q^n}p_n(x)p_n(y)\end{aligned}$$

and therefore

$$\begin{aligned}\Pi &= \prod_{n\geq 1}\exp\left(\frac{1}{n}\frac{1-t^n}{1-q^n}p_n(x)p_n(y)\right)\\ &= \prod_{n\geq 1}\sum_{m_n=0}^{\infty}\frac{1}{m_n!}\left(\frac{1}{n}\frac{1-t^n}{1-q^n}p_n(x)p_n(y)\right)^{m_n}\end{aligned}$$

in which the coefficient of $p_\lambda(x)p_\lambda(y)$, where $\lambda = (1^{m_1}2^{m_2}\dots)$, is seen to be $z_\lambda(q,t)^{-1}$. □

(11.6) *For each integer $n \geq 0$ let $(u_\lambda), (v_\lambda)$ be F-bases of Λ_F^n, indexed by the partitions of n. Then the following conditions are equivalent*:

(a) $\langle u_\lambda, v_\mu\rangle_{q,t} = \delta_{\lambda\mu}$ *for all* λ, μ;

(b) $\sum_\lambda u_\lambda(x)v_\lambda(y) = \Pi(x,y;q,t)$.

The proof follows the same lines as that of (6.5).

We shall now show how to construct the symmetric functions $P_\lambda(q,t)$. For this purpose we shall construct an F-linear operator

$$E = E_{q,t} : \Lambda_F \to \Lambda_F \tag{11.7}$$

having the following properties:

$$Em_\lambda = \sum_{\mu\leq\lambda} c_{\lambda\mu}m_\mu \tag{11.7.1}$$

for each partition λ, with coefficients $c_{\lambda\mu} \in F$;

$$\langle Ef, g\rangle_{q,t} = \langle f, Eg\rangle_{q,t} \tag{11.7.2}$$

for all $f, g \in \Lambda_F$;

$$c_{\lambda\lambda} \neq c_{\mu\mu} \quad \text{if } \lambda \neq \mu. \tag{11.7.3}$$

These three properties say respectively that the matrix of E relative to the basis (m_λ) is triangular (11.7.1); that E is selfadjoint (11.7.2); and that the eigenvalues of E are distinct (11.7.3).

The P_λ are then just the eigenfunctions (or eigenvectors) of the operator E:

(11.8) *For each partition λ there is a unique symmetric function $P_\lambda \in \Lambda_F$ satisfying*

$$P_\lambda = \sum_{\mu\leq\lambda} u_{\lambda\mu}m_\mu \tag{11.8.1}$$

with $u_{\lambda\mu} \in F$ and $u_{\lambda\lambda} = 1$;

$$EP_\lambda = c_{\lambda\lambda}P_\lambda. \tag{11.8.2}$$

PROOF. From (11.8.1) and (11.7.1) we have

$$EP_\lambda = \sum_{\mu \le \lambda} u_{\lambda\mu} E m_\mu = \sum_{\nu \le \mu \le \lambda} u_{\lambda\mu} c_{\mu\nu} m_\nu$$

and

$$c_{\lambda\lambda} P_\lambda = \sum_{\nu \le \lambda} c_{\lambda\lambda} u_{\lambda\nu} m_\nu$$

so that (11.8.1) and (11.8.2) are satisfied if and only if

$$c_{\lambda\lambda} u_{\lambda\nu} = \sum_{\nu \le \mu \le \lambda} u_{\lambda\mu} c_{\mu\nu}$$

that is to say, if and only if

$$(c_{\lambda\lambda} - c_{\nu\nu}) u_{\lambda\nu} = \sum_{\nu < \mu \le \lambda} u_{\lambda\mu} c_{\mu\nu}$$

whenever $\nu < \lambda$. Since $c_{\lambda\lambda} \ne c_{\nu\nu}$ by (11.7.3), this relation determines $u_{\lambda\mu}$ uniquely in terms of the $u_{\lambda\mu}$ such that $\nu < \mu \le \lambda$. Hence the coefficients $u_{\lambda\mu}$ in (11.8.1) are uniquely determined, given that $u_{\lambda\lambda} = 1$. □

With P_λ as defined in (11.8) we have, by the selfadjointness of E,

$$c_{\lambda\lambda}\langle P_\lambda, P_\mu\rangle_{q,t} = \langle EP_\lambda, P_\mu\rangle_{q,t} = \langle P_\lambda, EP_\mu\rangle_{q,t} = c_{\mu\mu}\langle P_\lambda, P_\mu\rangle_{q,t}$$

and by (11.7.3) we conclude that $\langle P_\lambda, P_\mu\rangle_{q,t} = 0$ if $\lambda \ne \mu$, i.e., the P_λ satisfy (11.1) and (11.2).

It remains therefore to construct an operator E satisfying (11.7.1)–(11.7.3). For this purpose we shall work initially in $\Lambda_{n,F} = F[x_1, \ldots, x_n]^{S_n}$. For any polynomial $f(x_1, \ldots, x_n)$, symmetric or not, we define

$$(T_{q,x_i} f)(x_1, \ldots, x_n) = f(x_1, \ldots, qx_i, \ldots, x_n),$$
$$(T_{t,x_i} f)(x_1, \ldots, x_n) = f(x_1, \ldots, tx_i, \ldots, x_n),$$

for $1 \le i \le n$. Now define an operator D_n by

(11.9)
$$D_n = a_\delta^{-1} \sum_{i=1}^{n} (T_{t,x_i} a_\delta) T_{q,x_i} = \sum_{i=1}^{n} \left(\prod_{j \ne i} \frac{tx_i - x_j}{x_i - x_j} \right) T_{q,x_i}.$$

A more useful expression for D_n is

(11.9′)
$$D_n = a_\delta^{-1} \sum_{w \in s_n} \varepsilon(w) \sum_{i=1}^{n} t^{(w\delta)_i} x^{w\delta} T_{q,x_i}.$$

We shall first show that D_n maps $\Lambda_{n,F}$ into $\Lambda_{n,F}$, and more precisely that

(11.10) *For each partition λ of length $\le n$ we have*

$$D_n m_\lambda = \sum_\mu d_n(\mu) s_\mu$$

summed over all distinct derangements $\mu = (\mu_1, \dots, \mu_n)$ *of* λ, *where*

$$d_n(\mu) = \sum_{i=1}^{n} q^{\mu_i} t^{n-i}.$$

PROOF. We may write m_λ in the form

$$m_\lambda = c^{-1} \sum_{w_1 \in S_n} x^{w_1 \lambda},$$

where c is the order of the subgroup of S_n that fixes λ. From (11.9′) we have

$$cD_n m_\lambda = a_\delta^{-1} \sum_{w, w_1} \varepsilon(w) \sum_{i=1}^{n} q^{(w_1\lambda)_i} t^{(w\delta)_i} x^{w_1\lambda + w\delta}.$$

In this sum, w and w_1 run independently through S_n. If we put $w_2 = w^{-1} w_1$, we shall obtain

$$\begin{aligned} cD_n m_\lambda &= a_\delta^{-1} \sum_{w, w_2} \varepsilon(w) \left(\sum_{i=1}^{n} q^{(w_2\lambda)_i} t^{n-i} \right) x^{w(w_2\lambda + \delta)} \\ &= \sum_{w_2 \in S_n} d_n(w_2\lambda) s_{w_2\lambda}, \end{aligned}$$

which gives the result, since each derangement μ of λ appears c times in this sum. □

The operators D_n, for varying n, are not compatible with the restriction homomorphism $\Lambda_{n+1,F} \to \Lambda_{n,F}$ (1.1). Indeed, we have

$$d_{n+1}(\mu) = \sum_{i=1}^{n+1} q^{\mu_i} t^{n+1-i},$$

so that

$$d_{n+1}(\mu) = t d_n(\mu) + 1. \tag{11.11}$$

We therefore modify D_n as follows: let

$$E_n = t^{-n}(1 + (t-1)D_n)$$

so that by (11.10)

$$E_n m_\lambda = \sum_\mu e_n(\mu) s_\mu \tag{11.12}$$

summed over derangements μ of λ as before, where

$$e_n(\mu) = t^{-n}(1 + (t-1)d_n(\mu))$$

which by (11.11) is equal to $e_{n+1}(\mu)$. It follows that the operators E_n are compatible with the homomorphisms $\Lambda_{n+1,F} \to \Lambda_{n,F}$, and we define

$$E = E_{q,t} = \varprojlim_n E_n : \Lambda_F \to \Lambda_F.$$

It remains to verify that E satisfies the conditions (11.7.1)–(11.7.3). Of these, (11.7.1) follows from (11.12) and (7.2), and (11.7.3) is clear, since $d_n(\lambda) \neq d_n(\nu)$ if λ, ν are distinct partitions. To establish (11.7.2) we shall make use of the following criterion, which is an easy consequence of (11.6):

(11.13) *E is selfadjoint if and only if $E_x\Pi = E_y\Pi$, where $\Pi = \Pi(x,y;q,t)$, and E_x* (resp. E_y) *denotes E acting on symmetric functions of the x's (resp. the y's).*

Now the definition of Π (11.4) shows that

$$\Pi^{-1}T_{q,x_i}\Pi = \prod_j \frac{1-x_iy_j}{1-tx_iy_j}$$

which is independent of q. Hence $\Pi^{-1}E_x\Pi$ is independent of q, and we may therefore assume that $q = t$: so by (11.13) we are reduced to proving that E is selfadjoint when $q = t$, and the scalar product is that originally defined in §6, for which the Schur functions s_λ form an orthonormal basis (7.4). Now when $q = t$ we have

$$D_n = a_\delta^{-1} \circ \left(\sum_{i=1}^n T_{t,x_i}\right) \circ a_\delta$$

in Λ_n, so that

$$\begin{aligned} D_n s_\lambda &= a_\delta^{-1}\left(\sum_{i=1}^n T_{t,x_i}\right) a_{\lambda+\delta} \\ &= \left(\sum_{i=1}^n t^{\lambda_i+n-i}\right) s_\lambda. \end{aligned}$$

It follows that the s_λ are simultaneous eigenfunctions of E when $q = t$, and hence that E is selfadjoint. This completes the proof of (11.7.2), and hence of the existence of the symmetric functions $P_\lambda(q,t)$ satisfying (11.1) and (11.2). □

REMARK. For each partition λ let

$$f_\lambda(q,t) = (1-q)(1-t)\sum t^{i-1}q^{j-1}$$

summed over all (i,j) in the diagram of λ. Then

(11.14) $$EP_\lambda(q,t) = (1 - f_\lambda(q,t^{-1}))P_\lambda(q,t).$$

PROOF. Let $n \geq l(\lambda)$. Then

$$\begin{aligned} f_\lambda(q,t) &= (1-q)(1-t)\sum_{i=1}^n t^{i-1}(1+q+\cdots+q^{\lambda_i-1}) \\ &= (1-t)\sum_{i=1}^n t^{i-1}(1-q^{\lambda_i}) \\ &= 1-t^n+(1-t^{-1})\sum_{i=1}^n q^{\lambda_i}t^i \end{aligned}$$

so that

$$\begin{aligned} 1 - f_\lambda(q,t^{-1}) &= t^{-n} + (t-1)\sum_{i=1}^n q^{\lambda_i}t^{-i} \\ &= t^{-n}(1+(t-1)d_n(\lambda)) = e_n(\lambda). \end{aligned}$$ □

12. Further properties of the $P_\lambda(q,t)$

In this section we shall survey, mostly without proofs, some of the properties of the symmetric functions $P_\lambda(q,t)$.

(a) *Duality.*

The $P_\lambda(q,t)$ form an orthogonal basis of Λ_F, relative to the scalar product (11.3). Let $(Q_\lambda(q,t))$ be the dual basis, so that

$$\langle P_\lambda, Q_\mu\rangle_{q,t} = \delta_{\lambda\mu}$$

or equivalently (11.6)

$$\sum_\lambda P_\lambda(x;q,t)Q_\lambda(y;q,t) = \Pi(x,y;q,t). \tag{12.1}$$

Each Q_λ is a scalar multiple of P_λ, say $Q_\lambda = b_\lambda P_\lambda$, where

$$b_\lambda = b_\lambda(q,t) = \langle P_\lambda, P_\lambda\rangle_{q,t}^{-1}. \tag{12.2}$$

The Schur functions s_λ satisfy the relation $\omega(s_\lambda) = s_{\lambda'}$ relative to the involution ω, as we showed in §7 by comparing the determinantal formulas (7.9) and (7.10). There is an analogous statement for the $P_\lambda(q,t)$, which goes as follows. Define an F-algebra automorphism $\omega_{q,t}$ of Λ_F by

$$\omega_{q,t}(p_r) = (-1)^{r-1}\frac{1-q^r}{1-t^r}p_r$$

for each $r \geq 1$. Then for each partition λ we have

$$\omega_{q,t}P_\lambda(q,t) = Q_{\lambda'}(t,q). \tag{12.3}$$

For a proof of (12.3), see either [**M5**], §3, or [**M6**], Chapter VI, §5.

(b) *Specialization.*

Let u be a new indeterminate and define a homomorphism (or specialization)

$$\varepsilon_{u,t} : \Lambda_F \to F[u]$$

by

$$\varepsilon_{u,t}(p_r) = \frac{1-u^r}{1-t^r}$$

for each $r \geq 1$. If $u = t^n$, where n is a positive integer, we have

$$\begin{aligned}\varepsilon_{t^n,t}(p_r) &= \frac{1-t^{nr}}{1-t^r}\\ &= 1 + t^r + \cdots + t^{(n-1)r}\\ &= p_r(1,t,\ldots,t^{n-1})\end{aligned}$$

and therefore

$$\varepsilon_{t^n,t}(f) = f(1,t,\ldots,t^{n-1})$$

for any symmetric function f: i.e., the effect of $\varepsilon_{u,t}$ when $u = t^n$ is to evaluate at $(x_1,\ldots,x_n,x_{n+1},\ldots) = (1,t,\ldots,t^n,0,0,\ldots)$.

There is a closed formula for $\varepsilon_{u,t}(P_\lambda(q,t))$. In order to state it in a convenient form, we introduce the following notation. For each square $s=(i,j)$ in the diagram of a partition $\lambda=(\lambda_1,\lambda_2,\dots)$ let

$$\begin{aligned} a(s) &= \lambda_i - j, & a'(s) &= j-1, \\ l(s) &= \lambda'_j - i, & l'(s) &= i-1, \end{aligned}$$

so that $l'(s), l(s), a(s)$, and $a'(s)$ are respectively the numbers of squares in the diagram of λ to the north, south, east and west of the square s. The numbers $a(s)$ and $a'(s)$ may be called respectively the *arm-length* and the *arm-colength* of s, and $l(s), l'(s)$ the *leg-length* and *leg-colength*. The *hook-length* at s is $a(s)+l(s)+1$. With this explained, we have

$$\varepsilon_{u,t}(P_\lambda(q.t)) = \prod_{s\in\lambda} \frac{q^{a'(s)}u - t^{l'(s)}}{q^{a(s)}t^{l(s)+1}-1} \tag{12.4}$$

for each partition λ, the product on the right being over the squares s in the diagram of λ ([**M5**], §5; [**M6**], Ch. VI, §6).

(c) *Norms.*

Given (12.3) and (12.4) it is a straightforward matter to calculate the squared norm $\langle P_\lambda, P_\lambda\rangle_{q,t}$ for any partition λ. For this purpose we require

(12.5) *Let $f \in \Lambda_F$ be homogeneous of degree r. Then*

$$\varepsilon_{u,t}\omega_{t,q}(f) = (-q)^{-r}\varepsilon_{u,q^{-1}}(f).$$

Since both $\varepsilon_{u,t}$ and $\omega_{t,q}$ are F-algebra homomorphisms, it is enough to verify (12.5) when $f=p_r$, in which case it follows directly from the definitions. We now calculate, using (12.5) and (12.3):

$$\begin{aligned} \varepsilon_{u,t}P_\lambda(q,t) &= \varepsilon_{u,t}\omega_{t,q}Q_{\lambda'}(t,q) \\ &= (-q)^{-|\lambda|}\varepsilon_{u,q^{-1}}Q_{\lambda'}(t,q) \\ &= (-q)^{-|\lambda|}b_{\lambda'}(t,q)\varepsilon_{u,q^{-1}}P_{\lambda'}(t^{-1},q^{-1}) \end{aligned}$$

(since $P_{\lambda'}(t,q) = P_{\lambda'}(t^{-1},q^{-1})$). Hence, on replacing (λ,q,t) by (λ',t,q) we obtain

$$b_\lambda(q,t) = (-t)^{|\lambda|}\frac{\varepsilon_{u,q}P_{\lambda'}(t,q)}{\varepsilon_{u,t^{-1}}P_\lambda(q^{-1},t^{-1})}$$

which by (12.4) reduces to

$$b_\lambda(q,t) = \prod_{s\in\lambda}\frac{1-q^{a(s)}t^{l(s)+1}}{1-q^{a(s)+1}t^{l(s)}},$$

or equivalently, by (12.2),

$$\langle P_\lambda, P_\lambda\rangle_{q,t} = \prod_{s\in\lambda}\frac{1-q^{a(s)+1}t^{l(s)}}{1-q^{a(s)}t^{l(s)+1}}. \tag{12.6}$$

(d) *Symmetry.*

Here we shall work in $\Lambda_{n,F} = F[x_1,\dots,x_n]^{S_n}$. For each $f\in\Lambda_{n,F}$ and each partition $\lambda=(\lambda_1,\dots,\lambda_n)$ of length $\le n$ we define

$$v_\lambda(f) = f(q^{\lambda_1}t^{n-1}, q^{\lambda_2}t^{n-2},\dots,q^{\lambda_n})$$

so that in particular v_0 is the specialization previously denoted by $\varepsilon_{t^n,t}$. We now renormalize the P_λ as follows:

$$\tilde{P}_\lambda = P_\lambda / v_0(P_\lambda),$$

for each partition λ of length $\leq n$, so that $\tilde{P}_\lambda(t^{n-1}, t^{n-2}, \ldots, 1) - 1$. Then we have

(12.7) $$v_\lambda(\tilde{P}_\mu) = v_\mu(\tilde{P}_\lambda)$$

for any two partitions λ, μ *of length* $\leq n$ ([**M6**], Ch. VI, §6).

(e) *Another scalar product.*

Again we shall work in $\Lambda_{n,F}$, and we shall also assume, for simplicity of exposition, that $t = q^k$ where k is a nonnegative integer. Let

$$L_n = F[x_1^{\pm 1}, \ldots, x_n^{\pm 1}]$$

denote the F-algebra of Laurent polynomials in $x_1, \ldots, x_n$, i.e., of polynomials in the x_i and x_i^{-1}. If $f \in L_n$ let $\overline{f} = f(x_1^{-1}, \ldots, x_n^{-1})$ and let $[f]_1$ denote the constant term of f. Moreover let

(12.8) $$\Delta = \Delta(x; q, t) = \prod_{i \neq j} \frac{(x_i x_j^{-1}; q)_\infty}{(t x_i x_j^{-1}; q)_\infty}$$

$$= \prod_{i \neq j} \prod_{r=0}^{k-1} (1 - q^r x_i x_j^{-1})$$

so that $\Delta \in L_n$.

For $f, g \in \Lambda_{n,F}$ we now define

(12.9) $$\langle f, g \rangle'_n = \frac{1}{n!} [f \overline{g} \Delta]_1.$$

This is a symmetric, positive definite scalar product on $\Lambda_{n,F}$, and it is not difficult to show that the operator D_n (11.9) is selfadjoint relative to this scalar product. From this it follows, just as in §11, that the $P_\lambda(q,t)$ are pairwise orthogonal:

(12.10) *Let* λ, μ *be partitions of length* $\leq n$. *Then*

$$\langle P_\lambda, P_\mu \rangle'_n = 0$$

if $\lambda \neq \mu$.

It remains to calculate the scalar product $\langle P_\lambda, P_\lambda \rangle'_n$. One form of the answer is

(12.11) $$\langle P_\lambda, Q_\lambda \rangle'_n = c_n \prod_{s \in \lambda} \frac{1 - q^{a'(s)} t^{n - l'(s)}}{1 - q^{a'(s)+1} t^{n - l'(s) - 1}}$$

for a partition λ of length $\leq n$, where

$$c_n = \langle 1, 1 \rangle'_n = \frac{1}{n!} [\Delta]_1$$

is (apart from the factor $1/n!$) the constant term of Δ. (12.11) may be proved by induction on λ, by removing one square at a time from the diagram of λ ([**M6**], Ch. VI, §9).

If we now renormalize this scalar product by defining

(12.12) $$\langle f, g \rangle''_n = c_n^{-1} \langle f, g \rangle'_n$$

for $f, g \in \Lambda_{n,F}$, so that $\langle 1, 1\rangle''_n = 1$, then the scalar product $\langle f, g\rangle_{q,t}$ (11.3) is the limit as $n \to \infty$ of the scalar product (12.12). To be precise:

(12.13) *Let $f, g \in \Lambda_F$ and let $\rho_n : \Lambda_F \to \Lambda_{n,F}$ be the canonical homomorphism (so that $\rho_n(x_i) = x_i$ if $i \leq n$, and $\rho_n(x_i) = 0$ if $i > n$). Then*

$$\lim_{n\to\infty} \langle \rho_n f, \rho_n g\rangle''_n = \langle f, g\rangle_{q,t}.$$

PROOF. It is enough to verify this when $f = P_\lambda$ and $g = Q_\mu$. If $\lambda \neq \mu$, both scalar products are zero, by (11.2) and (12.10). If $\lambda = \mu$, it is clear from (12.11) that $\langle P_\lambda, Q_\lambda\rangle''_n \to 1$ as $n \to \infty$. □

Finally, an equivalent version of the scalar product formula (12.11) is

$$\langle P_\lambda, P_\lambda\rangle'_n = \prod_{1\leq i<j\leq n} \prod_{r=1}^{k-1} \frac{1 - q^{\lambda_i - \lambda_j + r} t^{j-i}}{1 - q^{\lambda_i - \lambda_j - r} t^{j-i}} \tag{12.14}$$

for any partition $\lambda = (\lambda_1, \ldots, \lambda_n)$ of length $\leq n$, where $t = q^k$.

In the definition (12.8) of Δ, and in the second version (12.14) of the scalar product formula, the structure of the root system A_{n-1} is clearly visible. In the next chapter we shall show that analogues of the $P_\lambda(q,t)$ can be defined for any root system, and that many aspects of the theory sketched here have their counterparts in this more general setting.

CHAPTER 2

Orthogonal Polynomials

1. Introduction

The orthogonal polynomials that we shall construct in this chapter are Laurent polynomials in several variables. They depend rationally on two parameters q and t, and there is a family of them attached to each root system R. (When R is of type A, they are essentially the $P_\lambda(q,t)$ defined in Chapter I, §11.) For particular values of the parameters q and t, these polynomials reduce to objects familiar in representation theory:

(i) When $q = t$, they are independent of q and are the Weyl characters for the root system R.

(ii) When $q = 0$, they are (up to a scalar factor) the polynomials that give the values of zonal spherical functions on a semisimple p-adic Lie group G relative to a maximal compact subgroup K, such that the restricted root system of (G, K) is the dual root system $R^\vee$.

(iii) When q and t both tend to 1, in such a way that $(1-t)/(1-q)$ tends to a definite limit k, then (for certain values of k) our polynomials give the values of zonal spherical functions on a real (compact or noncompact) symmetric space G/K that arise from finite-dimensional spherical representations of G, that is to say representations having a nonzero K-fixed vector. Here the root system R is the restricted root system of (G, K), and the parameter k is half the root multiplicity (assumed to be the same for all restricted roots).

Thus these two-parameter families of orthogonal polynomials constitute a sort of bridge between commutative harmonic analysis on real symmetric spaces and on their p-adic analogues.

All this is in fact a simplified description. The general picture is more elaborate and involves parameters q_α and t_α for each root $\alpha \in R$, such that $q_\alpha = q_\beta$ and $t_\alpha = t_\beta$ when α and β are roots of the same length. The necessary modifications will be indicated later.

2. Root systems

Let V be a real vector space of finite dimension, endowed with a positive-definite symmetric bilinear form $\langle u, v\rangle$. For each nonzero $\alpha \in V$ let s_α denote the orthogonal reflection in the hyperplane through the origin perpendicular to α, so that

$$s_\alpha(v) = v - \langle v, \alpha^\vee\rangle \alpha \tag{2.1}$$

for $v \in V$, where $\alpha^\vee = 2\alpha/\langle \alpha, \alpha\rangle$.

A *root system* R in V is a finite nonempty set of nonzero vectors (called *roots*) that span V and are such that for each pair $\alpha, \beta \in R$ we have

$$\langle \alpha^\vee, \beta \rangle \in \mathbb{Z} \tag{2.2}$$

and

$$s_\alpha(\beta) \in R. \tag{2.3}$$

Thus each reflection s_α ($\alpha \in R$) permutes R, and the group of orthogonal transformations of V generated by the s_α is a finite group W called the *Weyl group* of R.

We may remark straightaway that the integrality condition (2.2) by itself is extremely restrictive. Let $\alpha, \beta \in R$ and let θ be the angle between the vectors α and β. Then

$$4\cos^2\theta = \frac{4\langle \alpha, \beta\rangle^2}{\langle \alpha, \alpha\rangle\langle \beta, \beta\rangle} = \langle \alpha^\vee, \beta\rangle\langle \alpha, \beta^\vee\rangle,$$

hence is an *integer*, hence is $0, 1, 2, 3$, or 4. It follows that the only possibilities for θ are π/m or $\pi - (\pi/m)$, where $m = 1, 2, 3, 4$, or 6. (It also follows that $|\langle \alpha^\vee, \beta\rangle| \leq 4$, and hence any set of vectors in V satisfying (2.2) is necessarily finite.)

The vectors $\alpha^\vee$ for $\alpha \in R$ form a root system $R^\vee$, the *dual* of R.

If $\alpha \in R$, then $-\alpha \in R$ (because $-\alpha = s_\alpha(\alpha)$). The root system R is said to be *reduced* if for each $\alpha \in R$ the only scalar multiples of α that belong to R are $\pm\alpha$. Furthermore, R is said to be *irreducible* if it is not possible to partition R into two nonempty subsets such that each root in R_1 is orthogonal to each root in R_2 (which would imply that R_1 and R_2 are themselves root systems). Until further notice, we shall assume that R is both reduced and irreducible.

For those to whom these notations are unfamiliar, some examples to bear in mind are the following. Let $\varepsilon_1, \dots, \varepsilon_n$ be the standard basis of $\mathbb{R}^n$ ($n \geq 2$), with the usual scalar product, for which $\langle \varepsilon_i, \varepsilon_j\rangle = \delta_{ij}$. Then the vectors

$$\varepsilon_i - \varepsilon_j, \tag{A_{n-1}}$$

where $i \neq j$, form a root system (and V is the hyperplane in $\mathbb{R}^n$ perpendicular to $\varepsilon_1 + \cdots + \varepsilon_n$). The Weyl group is the symmetric group S_n, acting on V by permuting the ε_i.

Moreover, each of the sets of vectors

$$\pm\varepsilon_i \quad (1 \leq i \leq n), \qquad \pm\varepsilon_i \pm \varepsilon_j, \quad (1 \leq i < j \leq n), \tag{B_n}$$

$$\pm 2\varepsilon_i \quad (1 \leq i \leq n), \qquad \pm\varepsilon_i \pm \varepsilon_j \quad (1 \leq i < j \leq n), \tag{C_n}$$

$$\pm\varepsilon_i \pm \varepsilon_j \quad (1 \leq i < j \leq n) \tag{D_n}$$

is a root system. For B_n and C_n, the Weyl group is the group of all signed permutations of the ε_i, of order $2^n n!$ (the hyperoctahedral group). For D_n, it is a subgroup of index 2 of this group. The root systems B_n, C_n are duals of each other, and A_{n-1}, D_n are each self-dual.

In fact, the root systems $A_n (n \geq 1)$, B_n ($n \geq 2$), C_n ($n \geq 3$), and D_n ($n \geq 4$) almost exhaust the catalogue of reduced irreducible root systems (up to isomorphism). Apart from these, there are just five others, the "exceptional" root systems denoted by E_6, E_7, E_8, F_4, G_2. (In each case the numerical suffix is the dimension of the space V spanned by R, which is also called the *rank* of R.) I shall have little to say specifically about the exceptional root systems, but here for example is a simple description of E_8. Let $\mathbb{Z}^9$ be the integer lattice in $\mathbb{R}^9$, and let V be the hyperplane in $\mathbb{R}^9$ orthogonal to $\varepsilon_1 + \cdots + \varepsilon_9$. The orthogonal projection of $\mathbb{Z}^9$ on V is a lattice L in V, and $R = \{\alpha \in L : \langle \alpha, \alpha \rangle = 2\}$ is a root system of type E_8. There are 240 roots.

Now let R be any (reduced, irreducible) root system and let $v \in V$ be such that $\langle v, \alpha \rangle \neq 0$ for each $\alpha \in R$. Then the set R^+ of roots $\alpha \in R$ such that $\langle v, \alpha \rangle > 0$ is called a *system of positive roots* in R. Of course R^+ depends on the choice of the vector v, but it can be shown that any other system of positive roots is of the form wR^+ for a unique element $w \in W$; so there is really no loss of generality in fixing R^+ once and for all.

A root $\alpha \in R^+$ is *simple* if it is not the sum of two elements of R^+. Altogether there are $n = \dim V$ simple roots, say $\alpha_1, \ldots, \alpha_n$, and every $\alpha \in R^+$ is of the form $\sum_{i=1}^n m_i \alpha_i$, where the coefficients m_i are nonnegative integers. In A_{n-1}, for example, we may take R^+ to consist of the roots $\varepsilon_i - \varepsilon_j$ such that $i < j$, and then the simple roots are $\varepsilon_i - \varepsilon_{i+1}$ $(1 \leq i \leq n-1)$.

The abelian group Q generated by R, whose elements are the integral linear combinations of the roots, is a lattice in V (i.e., a free abelian group of rank $n = \dim V$) called the *root lattice.* Clearly the simple roots of $\alpha_1, \ldots, \alpha_n$ form a basis of Q. We denote by Q^+ the subsemigroup of Q consisting of all sums $\sum m_i \alpha_i$ where the coefficients are nonnegative integers.

Next, the set P of all $\lambda \in V$ such that $\langle \lambda, \alpha^\vee \rangle \in \mathbb{Z}$ for all $\alpha \in R$ is another lattice in V, called the *weight lattice.* We denote by P^+ the set of *dominant* weights $\lambda \in P$ such that $\langle \lambda, \alpha^\vee \rangle \in \mathbb{N}$ for all $\alpha \in R^+$. We have $P \supset Q$ (by (2.2)) but $P^+ \not\supset Q^+$ (unless $n = 1$, i.e., $R = A_1$). The quotient P/Q is a finite group, since both P and Q are lattices of the same rank n. Clearly both P and Q are stable under the action of the Weyl group W, and W acts trivially on P/Q. Each W-orbit in P contains exactly one dominant weight, i.e., P^+ is a fundamental region for the action of W on P.

When R is A_{n-1} and the simple roots are $\varepsilon_i - \varepsilon_{i+1}$ $(1 \leq i \leq n-1)$ as described above, the dominant weights are vectors of the form $\lambda = \sum \lambda_i \varepsilon_i$ where the λ_i are rational numbers such that the successive differences $\lambda_i - \lambda_{i+1}$ $(1 \leq i \leq n-1)$ are nonnegative integers, and $\sum \lambda_i = 0$. Thus each partition $\mu = (\mu_1, \ldots, \mu_n)$ of length $\leq n$ determines a dominant weight λ by the rule $\lambda_i = \mu_i - n^{-1}|\mu|$ $(1 \leq i \leq n)$, and two partitions μ, ν determine the same dominant weight if and only if $\mu_1 - \nu_1 = \mu_2 - \nu_2 = \cdots = \mu_n - \nu_n$.

3. Orbit sums and Weyl characters

Let F be a field of characteristic zero and let $A = F[P]$ be the group algebra over F of the lattice P. Since the group operation in P is addition, we shall use an exponential notation in A, and denote by e^λ the element of A corresponding to $\lambda \in P$. These "formal exponentials" e^λ form an F-basis of A, such that $e^\lambda e^\mu = e^{\lambda+\mu}$ and $(e^\lambda)^{-1} = e^{-\lambda}$. In particular, $e^0 = 1$ is the identity element of A.

The Weyl group W acts on P and therefore also on A: $w(e^\lambda) = e^{w\lambda}$ for $\lambda \in P$ and $w \in W$. We denote by A^W the subalgebra of W-invariant elements in A.

Since each W-orbit in P meets P^+ exactly once, it follows that the *orbit-sums*

$$m_\lambda = \sum_{\mu \in W\lambda} e^\mu, \tag{3.1}$$

where $\lambda \in P^+$ and $W\lambda$ is the W-orbit of λ, form an F-basis of A^W.

Another F-basis of A^W is obtained as follows. Let

$$\rho = \frac{1}{2} \sum_{\alpha \in R^+} \alpha \tag{3.2}$$

and define

$$\delta = \prod_{\alpha \in R^+} (e^{\alpha/2} - e^{-\alpha/2}). \tag{3.3}$$

In fact, $\rho \in P^+$ and $\delta \in A$: we have

$$\delta = \sum_{w \in W} \varepsilon(w) e^{w\rho} \tag{3.4}$$

where $\varepsilon(w) = \det(w) = \pm 1$. Thus δ is skew-symmetric for W, i.e., $w\delta = \varepsilon(w)\delta$ for $w \in W$. For each $\lambda \in P^+$, the sum

$$\sum_{w \in W} \varepsilon(w) e^{w(\lambda+\rho)}$$

is likewise skew-symmetric, and is divisible by δ in A; the quotient

$$\chi_\lambda = \delta^{-1} \sum_{w \in W} \varepsilon(w) e^{w(\lambda+\rho)} \tag{3.5}$$

is an element of A^W called the *Weyl character* corresponding to λ. In terms of the orbit-sums we have

$$\chi_\lambda = m_\lambda + \sum_{\mu < \lambda} K_{\lambda\mu} m_\mu \tag{3.6}$$

where the $K_{\lambda\mu}$ are integers (indeed positive integers) and $\mu < \lambda$ means that $\lambda - \mu \in Q^+$ and $\lambda \neq \mu$. From (3.6) it follows that the χ_λ, $\lambda \in P^+$, from another F-basis of A^W.

(3.7) When R is A_{n-1} and the simple roots are $\varepsilon_i - \varepsilon_{i+1}$ $(1 \leq i \leq n-1)$ as in §2, let $\varphi_i = \varepsilon_i - \frac{1}{n}(\varepsilon_1 + \cdots + \varepsilon_n)$ and put $x_i = e^{\varphi_i}$ $(1 \leq i \leq n)$, so that $x_1 \cdots x_n = 1$. Then the orbit sums in this case are just the monomial symmetric functions (in n variables $x_1, \ldots, x_n$ subject to $x_1 \cdots x_n = 1$), δ is the Vandermonde determinant, and the Weyl characters are the Schur functions (compare (3.5) with Chapter I, (7.1)).

4. Scalar product

Let $q, t \in F$ and define

$$\Delta = \Delta(q,t) = \prod_{\alpha \in R} (e^\alpha; q)_\infty / (te^\alpha; q)_\infty \tag{4.1}$$

where (as in Chapter I, §11)

$$(a;q)_\infty = \prod_{r=0}^{\infty} (1 - aq^r).$$

Suppose first that $t = q^k$ where k is a nonnegative integer. Then

$$\Delta = \prod_{\alpha \in R} \prod_{r=0}^{k-1} (1 - q^r e^\alpha), \tag{4.2}$$

an element of A^W. We shall use Δ to define a (symmetric, nondegenerate) scalar product on A, as follows. If $f \in A$, say

$$f = \sum_{\lambda \in P} f_\lambda e^\lambda$$

let

$$\overline{f} = \sum_{\lambda \in P} f_\lambda e^{-\lambda}$$

and let $[f]_1$ denote the *constant term* f_0 of f. We then define the scalar product of f and g to be

$$\langle f, g \rangle = \frac{1}{|W|}[f\overline{g}\Delta]_1. \tag{4.3}$$

(If t is not a power of q, a different definition is needed. We take F to be the field of real numbers, $q \in (0,1)$ and $t \geq 0$. Let T be the torus $V/Q^\vee$, where $Q^\vee$ is the root lattice of $R^\vee$; the character group of T may be identified with the weight lattice P, and we may regard each e^λ, $\lambda \in P$, as a character of T by the rule

$$e^\lambda(\dot{x}) = \exp 2\pi i \langle \lambda, x \rangle$$

where $\dot{x} \in T$ is the image of $x \in V$, and exp is the exponential function. Then $f\overline{g}\Delta$ is a continuous function on the torus T, and we define

$$\langle f, g \rangle = |W|^{-1} \int_T f\overline{g}\Delta \tag{4.4}$$

the integration being with respect to normalized Haar measure.)

When R is A_{n-1}, the e^α are $x_i x_j^{-1}$ $(1 \leq i, j \leq n,\ i \neq j)$ in the notation introduced at the end of §3, and the scalar product (4.3) is that defined in Chapter I, (12.9).

5. The polynomials P_λ

(5.1) *There is a unique F-basis $(P_\lambda)_{\lambda \in P^+}$ of A^W such that*

(i) $P_\lambda = m_\lambda + \sum_{\mu < \lambda} u_{\lambda\mu} m_\mu$

where the coefficients $u_{\lambda\mu}$ are rational functions of q and t;

(ii) $\langle P_\lambda, P_\mu \rangle = 0$ *if* $\lambda \neq \mu$.

It is easy to see that the P_λ, if they exist, are unique. Their existence, however, requires proof. If the partial order $\lambda > \mu$ on P^+ were a *total* ordering, existence would follow directly from the Gram-Schmidt orthogonalization process. But it isn't a total ordering (unless $R = A_1$) and we should therefore have to extend it to a total ordering before applying the Gram-Schmidt mechanism. Thus the content of (5.1) is that however we extend the partial order $\lambda > \mu$ to a total order, we always obtain the same basis.

We shall give a proof of (5.1) in §§6 and 7. Before doing so, let us look at some special cases.

(1) When $t = 1$, we have $\Delta = 1$ and P_λ is the orbit-sum m_λ.

(2) When $t = q$, we have

$$\Delta = \prod_{\alpha \in R} (1 - e^\alpha) = \delta\overline{\delta}$$

by (3.3), and it follows that P_λ is the Weyl character χ_λ.

(3) When $q \to 0$, t being arbitrary, we have

$$\Delta = \prod_{\alpha \in R} (1 - e^\alpha)/(1 - te^\alpha).$$

In this case there is an explicit formula for P_λ, namely

$$P_\lambda = W_\lambda(t)^{-1} \sum_{w \in W} w \left(e^\lambda \prod_{\alpha \in R^+} \frac{1 - te^{-\alpha}}{1 - e^{-\alpha}} \right)$$

where

$$W_\lambda(t) = \sum_{\substack{w \in W \\ w\lambda = \lambda}} t^{l(w)}$$

and $l(w)$ is the *length* of w, that is to say the number of positive roots $\alpha \in R^+$ such that $w\alpha \notin R^+$.

These polynomials occur in nature as the values of zonal spherical functions on a p-adic Lie group, when t^{-1} is a prime power.

(4) Let $t = q^k$, fix k (which need not be an integer) and let $q \to 1$ (so that $t \to 1$ also). Then in the limit we have

$$\Delta = \prod_{\alpha \in R} (1 - e^\alpha)^k.$$

In this limiting case the polynomials P_λ were defined by Heckman and Opdam ([**HO**], [**O1**]) who called them *Jacobi polynomials.*

For particular values of k these polynomials P_λ again occur in nature as zonal spherical functions, but this time on a real semisimple Lie group G, relative to a maximal compact subgroup.

(5) Finally, when R is A_{n-1} the P_λ are the symmetric functions $P_\lambda(q,t)$ of Chapter I, §11, restricted to n variables $x_1, \ldots, x_n$ such that $x_1 \cdots x_n = 1$.

6. Proof of the existence theorem

The existence theorem (5.1) will be a consequence of the following proposition:

(6.1) *There is a linear operator* $D : A^W \to A^W$ *with the following three properties*:

(a) $\langle Df, g\rangle = \langle f, Dg\rangle$ *for all* $f, g \in A^W$;

(b) *For each* $\lambda \in P^+$, Dm_λ *is of the form*

$$Dm_\lambda = \sum_{\mu \leq \lambda} c_{\lambda\mu} m_\mu;$$

(c) *if* $\lambda, \mu \in P^+$ *and* $\lambda \neq \mu$, *then* $c_{\lambda\lambda} \neq c_{\mu\mu}$.

Property (a) says that D is selfadjoint; (b) says that the matrix of D is triangular, relative to the basis (m_λ) of A^W; and (c) says that the eigenvalues of D are all distinct.

Given an operator D with these three properties, for each $\lambda \in P^+$ let P_λ be the eigenfunction of D with eigenvalue $c_{\lambda\lambda}$, normalized so that the coefficient of m_λ in

P_λ is equal to 1. Then the P_λ so defined satisfy condition (i) of (5.1), and since

$$\begin{aligned} c_{\lambda\lambda}\langle P_\lambda, P_\mu\rangle &= \langle DP_\lambda, P_\mu\rangle = \langle P_\lambda, DP_\mu\rangle \\ &= c_{\mu\mu}\langle P_\lambda, P_\mu\rangle \end{aligned}$$

it follows that $\langle P_\lambda, P_\mu\rangle = 0$ if $\lambda \neq \mu$, so that (ii) is satisfied.

We come now to the proof of (6.1). For each $x \in V$ let $T_x : A \to A$ denote the F-linear mapping defined by

$$T_x e^\lambda = q^{\langle \lambda, x\rangle} e^\lambda \tag{6.2}$$

where $\lambda \in P$.

(6.3)(i) *T_x is an F-algebra automorphism of A.*
(ii) *For all $f, g \in A$ we have*

$$[\overline{f} T_x g]_1 = [\overline{g} T_x f]_1.$$

PROOF. (i) is clear from the definition. For (ii) we may take $f = e^\lambda$ and $g = e^\mu$, and then both sides are zero if $\lambda \neq \mu$, and equal to $q^{\langle \lambda, x\rangle}$ if $\lambda = \mu$. □

Now let f be the order of P/Q. The situation is simplest in the case where $f > 1$. In that case there exist $f - 1$ vectors $\pi \neq 0$ in V such that $\langle \pi, \alpha\rangle = 0$ or 1 for each $\alpha \in R^+$. (They are the *minuscule weights* for the dual root system $R^\vee$.)

Let

$$\Delta_+ = \prod_{\alpha \in R^+} (e^\alpha; q)_\infty / (te^\alpha; q)_\infty \tag{6.4}$$

and assume that $t = q^k$ with k an integer ≥ 0. Let π as above be a minuscule weight for $R^\vee$, and define an operator D_π on A by

$$D_\pi f = \sum_{w \in W} w(\Delta_+^{-1} T_\pi(\Delta_+ f)). \tag{6.5}$$

We shall show that D_π satisfies conditions (a) and (b) of (6.1), and that a suitably chosen linear combination of the D_π's satisfies (c) as well. Let us first show that D_π is selfadjoint (on the assumption that it maps A into A, which we shall justify shortly). Since

$$D_\pi f = \sum_{w \in W} \frac{w(T_\pi(\Delta_+ f))}{w\Delta_+},$$

and since $\Delta = w\Delta = w\Delta_+ \cdot \overline{w\Delta_+}$ for each $w \in W$, we have

$$\begin{aligned} \langle D_\pi f, g\rangle &= |W|^{-1} \sum_{w \in W} [w(T_\pi(\Delta_+ f) \cdot \overline{\Delta_+ g})]_1 \\ &= [T_\pi(\Delta_+ f) \cdot \overline{\Delta_+ g}]_1 \end{aligned}$$

for $f, g \in A^W$. By (6.3)(ii) this is symmetrical in f and g, so that

$$\langle D_\pi f, g\rangle = \langle D_\pi g, f\rangle = \langle f, D_\pi g\rangle.$$

Next, to show that D_π maps A into A, we rewrite (6.5) in the form

$$D_\pi f = \sum_{w \in W} w(\Phi_\pi \cdot T_\pi(f)), \tag{6.5$'$}$$

where

$$\Phi_\pi = \Delta_+^{-1} T_\pi(\Delta_+).$$

From the definitions, for each $\alpha \in R^+$ we have

$$T_\pi e^\alpha = \begin{cases} q e^\alpha & \text{if } \langle \pi, \alpha \rangle = 1, \\ e^\alpha & \text{otherwise,} \end{cases}$$

and therefore

$$\begin{aligned} \Phi_\pi &= \prod_{\substack{\alpha \in R^+ \\ \langle \pi, \alpha \rangle = 1}} \frac{1 - t e^\alpha}{1 - e^\alpha} \\ &= \prod_{\alpha \in R^+} \frac{1 - t^{\langle \pi, \alpha \rangle} e^\alpha}{1 - e^\alpha} \\ &= t^{\langle \pi, 2\rho \rangle} \prod_{\alpha \in R^+} \frac{1 - t^{-\langle \pi, \alpha \rangle} e^{-\alpha}}{1 - e^{-\alpha}} \\ &= t^{\langle \pi, 2\rho \rangle} e^\rho \delta^{-1} \prod_{\alpha \in R^+} (1 - t^{-\langle \pi, \alpha \rangle} e^{-\alpha}). \end{aligned}$$

For each subset X of R^+, let $|X|$ denote the number of elements in X and let

$$\sigma(X) = \sum_{\alpha \in X} \alpha.$$

When we multiply out the product above, we shall obtain

$$\Phi_\pi = t^{\langle \pi, 2\rho \rangle} \delta^{-1} \sum_{X \subset R^+} \varphi_X(t) e^{\rho - \sigma(X)}, \tag{6.6}$$

where

$$\varphi_X(t) = (-1)^{|X|} t^{-\langle \pi, \sigma(X) \rangle}. \tag{6.7}$$

We can now calculate $D_\pi e^\mu$, where $\mu \in P$. Since $w(T_\pi e^\mu) = q^{\langle \pi, \mu \rangle} e^{w\mu}$, and since $w\delta = \varepsilon(w)\delta$ for each $w \in W$, we shall obtain from (6.5′) and (6.6)

$$\begin{aligned} D_\pi e^\mu &= \delta^{-1} q^{\langle \pi, \mu \rangle} t^{\langle \pi, 2\rho \rangle} \sum_{X \subset R^+} \varphi_X(t) \sum_{w \in W} \varepsilon(w) e^{w(\mu + \rho - \sigma(X))}, \\ &= q^{\langle \pi, \mu \rangle} t^{\langle \pi, 2\rho \rangle} \sum_X \varphi_X(t) \chi_{\mu - \sigma(X)}, \end{aligned}$$

where χ_λ is defined by (3.5) for *all* $\lambda \in P$. This calculation shows that D_π maps A into A^W.

Now let $\lambda \in P^+$. Then we have

$$\begin{aligned} D_\pi m_\lambda &= \sum_{\mu \in W\lambda} D_\pi m_\mu \\ &= t^{\langle \pi, 2\rho \rangle} \sum_{X \subset R^+} \varphi_X(t) \sum_{\mu \in W\lambda} q^{\langle \pi, \mu \rangle} \chi_{\mu - \sigma(X)}. \end{aligned} \tag{6.8}$$

In this sum, either $\chi_{\mu - \sigma(X)} = 0$ or there exist $w \in W$ and $\nu \in P^+$ such that

$$\nu + \rho = w(\mu + \rho - \sigma(X)), \tag{6.9}$$

in which case $\chi_{\mu-\sigma(X)} = \varepsilon(w)\chi_\nu$. But $\rho - \sigma(X)$ is of the form

$$\rho - \sigma(X) = \frac{1}{2} \sum_{\alpha \in R^+} \varepsilon_\alpha \alpha,$$

where each coefficient ε_α is ± 1; hence $w(\rho - \sigma(X))$ is of the same form, and therefore

(6.10) $$\omega(\rho - \sigma(X)) = \rho - \sigma(y)$$

for some subset Y of R^+. From (6.9) and (6.10) it follows that

(6.11) $$\nu = w\mu - \sigma(Y) \leq w\mu \leq \lambda$$

and hence that $D_\pi m_\lambda$ is a linear combination of the χ_ν such that $\nu \in P^+$ and $\nu \leq \lambda$. Hence by (3.6) we have

(6.12) $$D_\pi m_\lambda = \sum_{\nu \leq \lambda} c_{\lambda\nu}(\pi) m_\nu$$

which shows that D_π satisfies (6.1)(a).

We must now calculate the leading coefficient $c_{\lambda\lambda}(\pi)$, which is the coefficient of χ_λ in (6.8). From (6.11) it follows that $\nu = \lambda$ if and only if Y is empty and $w\mu = \lambda$, that is to say if and only if $\mu = w^{-1}\lambda$ and $w(\rho - \sigma(X)) = \rho$, or equivalently $\sigma(X) = \rho - w^{-1}\rho$. But this implies that $X = R^+ \cap -wR^+$, hence

$$c_{\lambda\lambda}(\pi) = t^{\langle \pi, 2\rho \rangle} \sum_{w \in W} \varepsilon(w) \varphi_{R^+ \cap -wR^+}(t) q^{\langle \pi, w^{-1}\lambda \rangle}$$

where by (6.7)

$$\varphi_{R^+ \cap -wR^+}(t) = \varepsilon(w) t^{\langle \pi, w^{-1}\rho - \rho \rangle}.$$

So we obtain

(6.13) $$c_{\lambda\lambda}(\pi) = t^{\langle \pi, \rho \rangle} \sum_{w \in W} q^{\langle w\pi, \lambda \rangle} t^{\langle w\pi, \rho \rangle}.$$

It remains to examine whether the eigenvalues $c_{\lambda\lambda}(\pi)$ of D_π are all distinct as λ runs through P^+, for a suitable choice of the minuscule weight π. It will appear that this is so in all cases except D_n, $n \geq 4$ (and of course excepting E_8, F_4, and G_2, where there is no minuscule weight).

For $x \in V$ and $r \geq 1$ let

$$p_r(x) = \sum_{w \in W} \langle x, w\pi \rangle^r.$$

The p_r are W-invariant polynomial functions on V, and we have

(6.14) *Suppose that R is not of type D_n, $n \geq 4$, and that if R is of type A_n the minuscule weight π is φ_1, in the notation of (3.7). Then the p_r generate the $\mathbb{R}$-algebra of W-invariant polynomial functions on V, and hence separate the W-orbits in V.*

This is easily verified for R of type A, B, or C. For the remaining cases (E_6, E_7), we refer to **[M8]**.

Assume now that the hypotheses of (6.14) are satisfied, and that $\lambda, \mu \in P^+$ are such that $c_{\lambda\lambda}(\pi) = c_{\mu\mu}(\pi)$, i.e., that

$$\sum_{w \in W} q^{\langle w\pi, \lambda \rangle} t^{\langle w\pi, \rho \rangle} = \sum_{w \in W} q^{\langle w\pi, \mu \rangle} t^{\langle w\pi, \rho \rangle}$$

identically in q and t. By operating on both sides with $(q\partial/\partial q)^r$ and then setting $q = t = 1$, we obtain $p_r(\lambda) = p_r(\mu)$ for all $r \geq 1$. Hence by (6.14) λ and μ are in the same W-orbit, and therefore $\lambda = \mu$. It follows that the eigenvalues $c_{\lambda\lambda}(\pi)$ of D_π are all distinct.

There remains the case where R is of type D_n, $n \geq 4$. With the notation of §2, we may take R^+ to consist of the roots $\varepsilon_i \pm \varepsilon_j$ where $i < j$. Then P^+ consists of the vectors $\lambda = \sum \lambda_i \varepsilon_i$ for which the λ_i are all integers or all half-integers, and $\lambda_1 \geq \cdots \geq \lambda_{n-1} \geq |\lambda_n|$. The weights

$$\pi_1 = \tfrac{1}{2}(\varepsilon_1 + \cdots + \varepsilon_n), \qquad \pi_2 = \pi_1 - \varepsilon_n$$

are both minuscule, and the W-orbit of π_1 (resp. π_2) consists of all sums $\frac{1}{2}\sum \pm\varepsilon_i$ containing an even (resp. odd) number of minus signs. Hence the formula (6.13) gives

$$c_{\lambda\lambda}(\pi_1) \pm c_{\lambda\lambda}(\pi_2) = n! \prod_{i=1}^{n} (q^{\lambda_i} t^{n-i} \pm q^{-\lambda_i}).$$

Choose an integer $N > \frac{1}{2}n(n-1)$. Then the eigenvalues of the operator

$$D = \frac{1}{n!}((t^N + 1)D_{\pi_1} + (t^N - 1)D_{\pi_2})$$

are from above

$$c_\lambda = t^N \prod_{i=1}^{n} (q^{\lambda_i} t^{n-i} + q^{-\lambda_i}) + \prod_{i=1}^{n} (q^{\lambda_i} t^{n-i} - q^{-\lambda_i}).$$

Now suppose that $\lambda, \mu \in P^+$ are such that $c_\lambda = c_\mu$. From our choice of N it follows that

$$\prod_{i=1}^{n} (q^{\lambda_i} t^{n-i} + q^{-\lambda_i}) = \prod_{i=1}^{n} (q^{\mu_i} t^{n-i} + q^{-\mu_i}),$$
$$\prod_{i=1}^{n} (q^{\lambda_i} t^{n-i} - q^{-\lambda_i}) = \prod_{i=1}^{n} (q^{\mu_i} t^{n-i} - q^{-\mu_i}),$$

and hence that $\lambda = \mu$. So the eigenvalues of D are distinct, and (6.1) is established for all root systems R such that $f > 1$ (i.e., $P > Q$).

7. Proof of the existence theorem, concluded

In the cases (E_8, F_4, G_2) where there is no minuscule weight available, another construction is needed. Each $\alpha \in R^+$ is of the form $\alpha = \sum m_i \alpha_i$, where the α_i are the simple roots and the coefficients m_i are nonnegative integers. The sum of the coefficients m_i is called the *height* of α. There is a unique root $\varphi \in R^+$ of maximal height. If $\pi = \varphi^\vee$ we have, for each $\alpha \in R^+$,

$$(7.1) \qquad \langle \pi, \alpha \rangle = \begin{cases} 0 \text{ or } 1 \text{ if } \alpha \neq \varphi, \\ 2 \text{ if } \alpha = \varphi. \end{cases}$$

Thus π just fails to be a minuscule weight.

As in §6, let

$$\Phi_\pi = (T_\pi \Delta_+)/\Delta_+$$

and define an operator E_π on A^W as follows:

$$E_\pi f = \sum_{w \in W} w(\Phi_\pi \cdot U_\pi f)$$

where $U_\pi = T_\pi - 1$.

Let us first show that E_π is selfadjoint. If $f, g \in A^W$ we have

$$\begin{aligned}
\langle E_\pi f, g \rangle &= |W|^{-1} \sum_{w \in W} [w((T_\pi \Delta_+)(U_\pi f)) \cdot \overline{w(\Delta_+ g)}]_1 \\
&= [(T_\pi \Delta_+)(T_\pi f - f)\overline{\Delta_+ g}]_1 \\
&= [T_\pi(\Delta_+ f)\overline{\Delta_+ g}]_1 - [(T_\pi \Delta_+)\overline{\Delta_+} \cdot f\overline{g}]_1 \\
&= A(f,g) - B(f,g),
\end{aligned}$$

say. We have $A(f,g) = A(g,f)$ by (6.3)(ii). As to $B(f,g)$, let $G = (T_\pi \Delta_+)\overline{\Delta}_+$ and let w_0 be the longest element of the Weyl group W, such that $w_0 R^+ = -R^+$. Then $w_0 \pi = -\pi$ and $w_0 \Delta_+ = \overline{\Delta}_+$, so that

$$w_0 G = (T_{-\pi} \overline{\Delta}_+)\Delta_+ = \overline{T_\pi \Delta_+} \cdot \Delta_+ = \overline{G}$$

and therefore

$$\begin{aligned}
B(f,g) &= [G f \overline{g}]_1 = [(w_0 G) f \overline{g}]_1 \\
&= [\overline{G} f \overline{g}]_1 = [G \overline{f} g]_1 = B(g,f).
\end{aligned}$$

It follows that

$$\langle E_\pi f, g \rangle = \langle E_\pi g, f \rangle = \langle f, E_\pi g \rangle \tag{7.2}$$

and hence that E_π is selfadjoint.

From (7.1) we have

$$\begin{aligned}
\Phi_\pi &= \frac{1 - te^\varphi}{1 - e^\varphi} \cdot \frac{1 - qte^\varphi}{1 - qe^\varphi} \prod_{\substack{\alpha \in R^+ \\ \alpha \neq \varphi}} \frac{1 - t^{\langle \pi, \alpha \rangle} e^\alpha}{1 - e^\alpha} \\
&= t^{2\langle \pi, \rho \rangle} \delta^{-1} e^\rho \frac{(1 - t^{-1}e^{-\varphi})(1 - q^{-1}t^{-1}e^{-\varphi})}{1 - q^{-1}e^{-\varphi}} \prod_{\substack{\alpha \in R^+ \\ \alpha \neq \varphi}} (1 - t^{-\langle \pi, \alpha \rangle} e^{-\alpha})
\end{aligned}$$

and therefore

$$\Phi_\pi = t^{2\langle \pi, \rho \rangle} \delta^{-1} \frac{(1 - t^{-1}e^{-\varphi})(1 - q^{-1}t^{-1}e^{-\varphi}}{1 - q^{-1}e^{-\varphi}} \sum_X \varphi_X(t) e^{\rho - \sigma(X)} \tag{7.3}$$

summed over all subsets X of R^+ such that $\varphi \notin X$, where as in §6

$$\sigma(X) = \sum_{\alpha \in X} \alpha$$

and

$$\varphi_X(t) = (-1)^{|X|} t^{-\langle \pi, \sigma(X) \rangle}. \tag{7.4}$$

Now let $\lambda \in P^+$ and let $\mu \in W\lambda$. Since $\pi = \varphi^\vee$ we have

$$U_\pi m_\lambda = \sum_{\mu \in W\lambda} (q^{\langle \mu, \varphi^\vee \rangle} - 1) e^\mu. \tag{7.5}$$

Any $\mu \in W\lambda$ such that $\langle \mu, \varphi^\vee \rangle = 0$ will contribute nothing to this sum. The remaining elements of the orbit $W\lambda$ fall into pairs $\{\mu, s_\varphi \mu\}$ where $\langle \mu, \varphi^\vee \rangle > 0$. Hence we may rewrite (7.5) in the form

$$U_\pi m_\lambda = \sum_{\substack{\mu \in W\lambda \\ \langle \mu, \varphi^\vee \rangle > 0}} (q^{\langle \mu, \varphi^\vee \rangle - 1}) e^\mu (1 - (qe^\varphi)^{-\langle \mu, \varphi^\vee \rangle})$$

from which it follows that

$$\frac{U_\pi m_\lambda}{1 - q^{-1} e^{-\varphi}} = \sum_{\substack{\mu \in W\lambda \\ \langle \mu, \varphi^\vee \rangle > 0}} (q^{\langle \mu, \varphi^\vee \rangle - 1}) \sum_{j=0}^{\langle \mu, \varphi^\vee \rangle - 1} q^{-j} e^{\mu - j\varphi}. \tag{7.6}$$

From (7.3) and (7.6) we obtain

$$\begin{aligned} \Phi_\pi \cdot U_\pi m_\lambda = & t^{2\langle \pi, \rho \rangle} \delta^{-1} \sum_{X, \mu} \varphi_X(t) e^{\rho - \sigma(X)} (1 - t^{-1} e^{-\varphi}) \\ & (1 - q^{-1} t^{-1} e^{-\varphi})(q^{\langle \mu, \varphi^\vee \rangle} - 1) \sum_{j=0}^{\langle \mu, \varphi^\vee \rangle - 1} q^{-j} e^{\mu - j\varphi}. \end{aligned} \tag{7.7}$$

This is a sum of terms of the form $a\delta^{-1}e^\eta$, $\eta \in P$. Since $\sum_{w \in W} w(\delta^{-1} e^\eta) \in A^W$, it follows from (7.7) that $E_\pi m_\lambda \in A^W$, and hence that E_π maps A^W into A^W.

Moreover, the terms $a\delta^{-1}e^\eta$ that occur in (7.7) are such that

$$\eta = \rho - \sigma(X) + \mu - j\varphi,$$

where $0 \leq j \leq \langle \mu, \varphi^\vee \rangle + 1$ and $\langle \mu, \varphi^\vee \rangle \geq 1$ and $\varphi \notin X$. If η is not regular (i.e., if $w\eta = \eta$ for some $w \neq 1$ in W), it will contribute nothing to (7.7). If on the other hand η is regular, then we have $w\eta = \xi + \rho$ for some $\xi \in P^+$ and some $w \in W$, so that

$$\xi + \rho = w(\rho - \sigma(X)) + w(\mu - j\varphi). \tag{7.8}$$

There are two cases to consider.

(i) Suppose that $w\varphi = \alpha \in R^+$. As in §6, we have

$$w(\rho - \sigma(X)) = \rho - \sigma(X)$$

for some subset Y of R^+, and hence

$$\xi = w\mu - j\alpha - \sigma(Y) \leq w\mu \leq \lambda. \tag{7.9}$$

Consequently each such term $a\delta^{-1}e^\eta$ in (7.7) contributes $\varepsilon(w) a \chi_\xi$, where $\xi \leq \lambda$, to $E_\pi m_\lambda$. Moreover we have equality in (7.9) if and only if $j = 0$, the subset Y is empty and $w\mu = \lambda$, i.e., $\mu = w^{-1}\lambda$ and $\sigma(X) = \rho - w^{-1}\rho$, so that $X = R^+ \cap -wR^+$. The coefficient a of $\delta^{-1}e^\eta = \delta^{-1}e^{w^{-1}(\lambda + \rho)}$ in (7.7) is then

$$\begin{aligned} a &= \varphi_{R^+ \cap -wR^+}(t) \cdot t^{2\langle \pi, \rho \rangle} (q^{\langle w\pi, \lambda \rangle} - 1) \\ &= \varepsilon(w) t^{\langle \pi + w\pi, \rho \rangle} (q^{\langle w\pi, \lambda \rangle} - 1) \end{aligned} \tag{7.10}$$

and the corresponding contribution to $E_\pi m_\lambda$ is

$$t^{\langle \pi + w\pi, \rho \rangle} (q^{\langle w\pi, \lambda \rangle} - 1) \chi_\lambda. \tag{7.11}$$

(ii) Suppose now that $w\varphi = -\alpha$, where $\alpha \in R^+$, and let $\nu = s_\varphi \mu = \mu - \langle \mu, \varphi^\vee \rangle \varphi$. Then $\mu - j\varphi = \nu - \varphi + j'\varphi$, where $j' = \langle \mu, \varphi^\vee \rangle + 1 - j$, so that $0 \leq j' \leq \langle \mu, \varphi^\vee \rangle + 1$. Hence (7.8) now takes the form

$$\xi + \rho = w(\rho - \sigma(X) - \varphi) + w(\nu + j'\varphi).$$

Since $\varphi \notin X$, we have $w(\rho - \sigma(X) - \varphi) = \rho - \sigma(Y)$ for some subset Y of R^+, and therefore

$$\xi = w\nu - \sigma(Y) - j'\alpha \leq w\nu \leq \lambda. \tag{7.12}$$

So again each term $a\delta^{-1}\eta$ in (7.7) contributes $\varepsilon(w)a\chi_\xi$, where $\xi \leq \lambda$, to $E_\pi m_\lambda$.

Moreover, we have equality in (7.12) if and only if $j' = 0$, the subset Y is empty, and $w\nu = \lambda$, i.e., $\nu = w^{-1}\lambda$ and $\sigma(X) + \varphi = \rho - w^{-1}\rho$, so that $X \cup \{\varphi\} = R^+ \cap -wR^+$ and $\mu = s_\varphi w^{-1}\lambda$ and $j = \langle \mu, \varphi^\vee \rangle + 1$. Hence the coefficient of $\delta^{-1}e^\eta = \delta^{-1}e^{w^{-1}(\lambda+\rho)}$ in $\Phi_\pi . U_\pi m_\lambda$ is now

$$a = \varphi_X(t) \cdot t^{2\langle \pi, \rho \rangle - 2} q^{-\langle \pi, \mu \rangle}(q^{\langle \pi, \mu \rangle} - 1)$$

where $X = (R^+ \cap -wR^+) - \{\varphi\}$. Since

$$\langle \pi, \mu \rangle = -\langle \pi, w^{-1}\lambda \rangle = -\langle w\pi, \lambda \rangle$$

and $\varphi_X(t) = -\varepsilon(w)t^{-\langle \pi, \rho - w^{-1}\rho \rangle + 2}$, it follows that a is given by the same formula (7.10) as before, and the corresponding contribution to $E_\pi m_\lambda$ is again given by (7.11).

To recapitulate, these calculations show that $E_\pi m_\lambda$ is a linear combination of the Weyl characters χ_ξ such that $\xi \in P^+$ and $\xi \leq \lambda$, and that the coefficient of χ_λ is

$$t^{\langle \pi, \rho \rangle} \sum_{w \in W} t^{\langle w\pi, \rho \rangle}(q^{\langle w\pi, \lambda \rangle} - 1). \tag{7.13}$$

As in §6, we conclude that

$$E_\pi m_\lambda = \sum_{\mu \leq \lambda} c_{\lambda\mu}(\pi) m_\mu$$

with leading coefficient $c_{\lambda\lambda}(\pi)$ given by (7.13).

To complete the proof of (5.1), it remains to establish that the $c_{\lambda\lambda}(\pi)$, $\lambda \in P^+$, are all distinct when R is of type E_8, F_4, or G_2. For this purpose we argue as in the last part of §6: if $\lambda, \mu \in P^+$ are such that $c_{\lambda\lambda}(\pi) = c_{\mu\mu}(\pi)$, then by operating with $(q\partial/\partial q)^r$ and then setting $q = t = 1$ we shall obtain

$$p_r(\lambda) = p_r(\mu)$$

for all $r \geq 1$, where

$$p_r(x) = \sum_{w \in W} \langle w\pi, x \rangle^r.$$

In each case $\pi = \varphi^\vee$ is proportional to the highest root of R, and it is known [**M8**] that when R is of type E_8, F_4, or G_2 the polynomial functions p_r, $r \geq 1$, generate the $\mathbb{R}$-algebra of W-invariant polynomial functions on V, and therefore separate the W-orbits in V. Hence λ and μ lie in the same W-orbit, and so $\lambda = \mu$, and the proof is complete. □

8. Some properties of the P_λ

Many of the properties of the symmetric polynomials $P_\lambda(x_1, \dots, x_n; q, t)$ surveyed in Chapter I, §12, namely the specialization formula (12.4), symmetry (12.7), and the second norm formula (12.14), generalize to arbitrary root systems.*

(a) *Norms*

If one has a family of orthogonal polynomials P_λ, the first thing one needs to know is the value of the scalar product $\langle P_\lambda, P_\lambda \rangle$. To state the result, we recall that the *q-gamma function* $\Gamma_q(x)$ is defined (for $x \in \mathbb{R}$, $x \notin -\mathbb{N}$) by

$$\Gamma_q(x) = \frac{(q;q)_\infty}{(q^x;q)_\infty}(1-q)^{1-x}.$$

(Here we regard q as a positive real number $\in (0,1)$ rather than an indeterminate.) It is convenient also to define

$$\Gamma_q^*(x) = 1/\Gamma_q(1-x).$$

Let $t = q^k$ and define, for $x \in V$,

$$c(x) = \prod_{\alpha \in R^+} \frac{\Gamma_q(\langle x, \alpha^\vee \rangle)}{\Gamma_q(\langle x, \alpha^\vee \rangle + k)}, \tag{8.1}$$

and

$$c^*(x) = \prod_{\alpha \in R^+} \frac{\Gamma_q^*(\langle x, \alpha^\vee \rangle)}{\Gamma_q^*(\langle x, \alpha^\vee \rangle + k)}. \tag{8.2}$$

Then we have

$$\langle P_\lambda, P_\lambda \rangle = \frac{c^*(-\lambda - k\rho)}{c(\lambda + k\rho)} \tag{8.3}$$

for all $\lambda \in P^+$.

If k is a nonnegative integer, an equivalent statement is

$$\langle P_\lambda, P_\lambda \rangle = \prod_{\alpha \in R^+} \prod_{i=1}^{k-1} \frac{1 - q^{\langle \lambda + k\rho, \alpha^\vee \rangle + i}}{1 - q^{\langle \lambda + k\rho, \alpha^\vee \rangle - i}}. \tag{8.3$'$}$$

Notice that when $\lambda = 0$ we have $P_\lambda = 1$, so that in this case (8.3′) gives the constant term of Δ, which was the subject of earlier conjectures ([**M3**], [**M9**]). In the limiting case $q \to 1$ (8.3) was first proved by Opdam [**O2**], and then in full generality by Cherednik [**C2**].

(b) *Specialization.*

Here it is convenient to regard each $f \in A$ as a function on V, as follows: if $x \in V$ and $f = \sum f_\lambda e^\lambda$, then

$$f(x) = \sum f_\lambda q^{\langle \lambda, x \rangle}.$$

Let

$$\rho^\vee = \frac{1}{2} \sum_{\alpha \in R^+} \alpha^\vee$$

---

*At the time (March 1993) these lectures were given, all this was conjectural.

(warning: $\rho^\vee \neq 2\rho/\langle\rho,\rho\rangle$). Then we have

$$P_\lambda(k\rho^\vee) = q^{-\langle\lambda,k\rho^\vee\rangle} c(k\rho)/c(\lambda+k\rho), \tag{8.4}$$

for all $\lambda \in P^+$.

When R is of type A_{n-1}, this formula is equivalent to that of Chapter I, (12.4) (with $u = t^n$).

In the limiting case $q \to 1$, (8.4) was proved by Opdam ([**O2**], Cor. 5.2), and then in full generality by Cherednik [**C3**].

(c) *Symmetry.*

For $\lambda \in P^+$ let

$$\widetilde{P}_\lambda = P_\lambda/P_\lambda(k\rho^\vee).$$

Then we have

$$\widetilde{P}_\lambda(\mu + k\rho^\vee) = \widetilde{P}_\mu(\lambda + k\rho) \tag{8.5}$$

for all $\lambda \in P^+$ and $\mu \in (P^\vee)^+$, where $P^\vee$ is the weight lattice of $R^\vee$, and on the right hand side of (8.5) P_μ is an orthogonal polynomial for $R^\vee$, so that $\widetilde{P}_\mu = P_\mu/P_\mu(k\rho)$.

When R is of type A_{n-1} (so that $R = R^\vee$), (8.5) becomes the formula (12.7) of Chapter I, and is due to Koornwinder. The general case is due to Cherednik [**C3**].

9. The general case

For ease of exposition we have presented a simplified version of the theory. In the more general version we take as starting point two irreducible root systems R and R' in the same vector space V, such that R' (but not necessarily R) is reduced, and such that R and R' have the same Weyl group W. In this situation, for each $\alpha \in R$ there exists a unique positive real number u_α such that $u_\alpha^{-1}\alpha \in R'$. We have $u_\alpha = u_\beta$ if α and β are in the same W-orbit, and the mapping $\alpha \mapsto u_\alpha^{-1}\alpha$ from R to R' is surjective.

Let q be a real number such that $0 \leq q < 1$, and for each $\alpha \in R$ let $q_\alpha = q^{u_\alpha}$. Moreover, for each $\alpha \in R$ let t_α be a real number ≥ 0, such that $t_\alpha = t_\beta$ if α and β are in the same W-orbit. The weight function Δ is now

$$\Delta = \prod_{\alpha\in R} (t_{2\alpha}^{1/2} e^\alpha; q_\alpha)_\infty / (t_\alpha t_{2\alpha}^{1/2} e^\alpha; q_\alpha)_\infty \tag{9.1}$$

where $t_{2\alpha} = 1$ if $2\alpha \notin R$, and the scalar product on A^W is defined by

$$\langle f, g\rangle = |W|^{-1}[f\bar{g}\Delta]_1$$

as before. With these modifications the existence theorem (5.1) is still valid, and the proof is the same, with the necessary cosmetic adjustments.

The definition of $c(\lambda)$ is now

$$c(\lambda) = \prod_{\alpha\in R^+} \frac{\Gamma_{q_\alpha}(\langle\lambda,\alpha^\vee\rangle + \frac{1}{2}k_{\alpha/2})}{\Gamma_{q_\alpha}(\langle\lambda,\alpha^\vee\rangle + \frac{1}{2}k_{\alpha/2} + k_\alpha)} \tag{9.2}$$

where k_α is defined by $t_\alpha = q_\alpha^{k_\alpha}$, and $k_{\alpha/2} = 0$ if $\frac{1}{2}\alpha \notin R$. With $c^*(\lambda)$ defined in the same way, by replacing each Γ by Γ^*, the value of $\langle P_\lambda, P_\lambda\rangle$ is

$$\langle P_\lambda, P_\lambda\rangle = \frac{c^*(-\lambda - \rho_k)}{c(\lambda + \rho_k)}, \tag{9.3}$$

where now

(9.4) $$\rho_k = \frac{1}{2} \sum_{\alpha \in R^+} k_\alpha \alpha.$$

Again, the specialization formula (8.4) now takes the form

(9.5) $$P_\lambda(\rho_k^*) = q^{-\langle \lambda, \rho_k^* \rangle} c(\rho_k)/c(\lambda + \rho_k)$$

where

(9.6) $$\rho_k^* = \frac{1}{2} \sum_{\alpha \in R^+} k_\alpha u_\alpha \alpha^\vee.$$

CHAPTER 3

Postscript

As explained in the preface, Chapters I and II constitute a somewhat expanded version of the oral lectures delivered in March 1993, with the exception that the formulas (8.3), (8.4), and (8.5) of Chapter II were there presented as conjectures. The aim of this postscript is to give a brief survey of developments since that date, and in particular to describe how the affine Hecke algebra (to be defined below) enters into the picture. For fuller details and proofs we refer the reader to the papers of Cherednik [**C1**]–[**C4**], and to the report [**M7**].

1. The affine root system and the extended affine Weyl group

We shall retain the notation of Chapter II, §2, so that R is a reduced irreducible root system spanning a real vector space V of dimension $n \geq 1$. However, we shall now denote the Weyl group of R by W_0 (instead of W). Let $Q^\vee$ denote the root lattice and $P^\vee$ the weight lattice of the dual root system $R^\vee$.

We shall regard each $\alpha \in R$ as a linear function on V: $\alpha(x) = \langle \alpha, x\rangle$ for $x \in V$. Also let c denote the constant function 1 on V. Then

$$S = S(R) = \{\alpha + nc : \alpha \in R, n \in \mathbb{Z}\} \tag{1.1}$$

is the *affine root system* associated with R. The elements of S are affine-linear functions on V, called *affine roots*, and we shall denote them by italic letters $a, b, \ldots$.

For each $a \in S$, let H_a denote the affine hyperplane in V on which a vanishes, and let s_a denote the orthogonal reflection in this hyperplane. The *affine Weyl group* W_S is the group of affine isometries of V generated by these reflections. For each $\alpha \in R$, the mapping $s_\alpha \circ s_{\alpha+c}$ takes $x \in V$ to $x + \alpha^\vee$, so that

$$\tau(\alpha^\vee) = s_\alpha \circ s_{\alpha+c}$$

is translation by $\alpha^\vee$. It follows that W_S contains a subgroup of translations isomorphic to $Q^\vee$, and we have

$$W_S = W_0 \ltimes \tau(Q^\vee) \tag{1.2}$$

(semidirect product).

The *extended affine Weyl group* is

$$W = W_0 \ltimes \tau(P^\vee). \tag{1.3}$$

It acts on V as a discrete group of isometries, and hence by transposition on functions on V. As such, it permutes the affine roots $a \in S$.

As in Chapter II, let R^+ be a system of positive roots in R and let $\alpha_1, \ldots, \alpha_n$ be the simple roots determined by R^+, and $\varphi \in R^+$ the highest root. Correspondingly, the affine roots $a_0, a_1, \ldots, a_n$ where

$$a_0 = -\varphi + c, \quad a_i = \alpha_i \quad (1 \leq i \leq n)$$

form a set of simple roots for S. Let

$$C = \{x \in V : a_i(x) \geq 0 \ (0 \leq i \leq n)\}$$

so that C is an open n-simplex bounded by the hyperplanes H_{a_i} $(0 \leq i \leq n)$. Then W_S is generated by the reflections $s_i = s_{a_i}$ $(0 \leq i \leq n)$, subject to the relations

$$s_i^2 = 1, \tag{1.4}$$

$$s_i s_j s_i \cdots = s_j s_i s_j \cdots \tag{1.5}$$

whenever $i \neq j$ and $s_i s_j$ has finite order m_{ij} in W_s, there being m_{ij} terms on either side of (1.5). In other words, W_S is a Coxeter group on the generators $s_0, s_1, \ldots, s_n$.

The connected components of $V - \cup_{a \in S} H_a$ are open simplexes, each congruent to C, and each component is of the form wC for a unique $w \in W_S$. Thus for example when R is of type A_2 we obtain the familiar tessellation of the Euclidean plane into congruent equilateral triangles.

An affine root $a \in S$ is *positive* (resp. *negative*) relative to C if $a(x) > 0$ (resp. $a(x) < 0$) for all $x \in C$. Let S^+ (resp. S^-) denote the set of positive (resp. negative) affine roots. Then $S^- = -S^+$, and $S = S^+ \cup S^-$. Explicitly, we have

$$S^+ = \{\alpha + (\chi(\alpha) + n)c : \alpha \in R, n \in \mathbb{N}\} \tag{1.6}$$

where $\chi(\alpha) = 0$ if $\alpha \in R^+$, and $\chi(\alpha) = 1$ if $\alpha \in R^- = -R^+$.

We now define a length function on the extended group W. If $w \in W$, let

$$l(w) = \operatorname{card}(S^+ \cap wS^-), \tag{1.7}$$

the number of positive affine roots made negative by w. Equivalently, $l(w)$ is the number of hyperplanes H_a, $a \in S$, that separate C from wC.

Now W, unlike W_S, is in general not a Coxeter group (unless $P^\vee = Q^\vee$) and many contain elements $\neq 1$ of length zero. Let

$$\Omega = \{w \in W : l(w) = 0\}. \tag{1.8}$$

The elements of Ω stabilize the simplex C, and hence permute the simple affine roots $a_0, a_1, \ldots, a_n$. For each $w \in W$ there exists a unique $w' \in W_S$ such that $wC = w'C$, and hence w factorizes uniquely as $w = w'v$ with $w' \in W_S$ and $v \in \Omega$. Consequently we have

$$W = W_S \rtimes \Omega \tag{1.9}$$

(semidirect product). From (1.2), (1.3), and (1.9) it follows that $\Omega \cong W/W_S \cong P^\vee/Q^\vee$, hence is a finite abelian group.

2. The braid group

The *braid group* B of W is the group with generators $T(w)$, $w \in W$, and relations

$$T(v)T(w) = T(vw) \tag{2.1}$$

whenever $l(vw) = l(v) + l(w)$. We shall denote $T(s_i) = T(s_{a_i})$ by T_i $(0 \leq i \leq n)$, and $T(\omega)$ $(\omega \in \Omega)$ simply by ω. Then B is generated by $T_0, T_1, \ldots, T_n$ and Ω subject to the following relations:

(a) the counterparts of (1.5), namely the *braid relations*

$$T_i T_j T_i \cdots = T_j T_i T_j \cdots \tag{2.2}$$

with m_{ij} terms on either side;

(b) the relations

$$\omega T_i \omega^{-1} = T_j \tag{2.3}$$

for $\omega \in \Omega$, where $\omega(a_i) = a_j$.

Let $\lambda \in (P^\vee)^+$ be a dominant weight for $R^\vee$, and define

$$Y^\lambda = T(\tau(\lambda)).$$

We have $l(\tau(\lambda)) = \langle \lambda, 2\rho \rangle$ if λ is dominant, and hence by (2.1)

$$Y^\lambda Y^\mu = Y^{\lambda+\mu} \tag{2.4}$$

if λ, μ are both dominant. If now λ is any element of $P^\vee$, we can write $\lambda = \mu - \nu$, with μ and ν both dominant, and we define

$$Y^\lambda = Y^\mu (Y^\nu)^{-1}. \tag{2.5}$$

In view of (2.4), this definition is unambiguous. The elements Y^λ, $\lambda \in P^\vee$, form a commutative subgroup of B, isomorphic to $P^\vee$.

Let λ_i $(1 \leq i \leq n)$ be the fundamental weights for $R^\vee$, defined by $\langle \lambda_i, \alpha_j \rangle = \delta_{ij}$ $(1 \leq i, j \leq n)$, and let $Y_i = Y^{\lambda_i}$. Then B is generated by $T_1, \ldots, T_n, Y_1, \ldots, Y_n$ subject to the following relations:

(a) the braid relations (2.2) not involving T_0;

(b) the relations $Y_i Y_j = Y_j Y_i$ $(1 \leq i, j \leq n)$,

(c) the relations

$$T_i Y_j = Y_j T_i \tag{2.6}$$

if $i \neq j$, and

$$Y_i = T_i Y_i Y^{-\alpha_i^\vee} T_i \tag{2.7}$$

for $1 \leq i \leq n$.

3. The affine Hecke algebra

Let m be the smallest positive even integer such that $mP \subset Q$, let q be an indeterminate and let $K = \mathbb{Q}(q^{1/m})$. Also, as in Chapter II, let $t = q^k$ where k is an integer ≥ 0. The *Hecke algebra* H of W is the quotient of the group algebra $K[B]$ of the braid group by the ideal generated by the elements $(T_i - t^{1/2})(T_i + t^{-1/2})$ $(0 \leq i \leq n)$. For each $w \in W$, we denote the image of $T(w)$ in H by the same symbol $T(w)$; these elements form a K-basis of H. Thus H is generated over K by $T_0, T_1, \ldots, T_n$ and Ω subject to the relations (2.2) and (2.3), together with

$$(T_i - t^{1/2})(T_i + t^{-1/2}) = 0 \tag{3.1}$$

for $0 \leq i \leq n$.

The following formula, due to Lusztig [**L**], is fundamental for what follows:

(3.2) *Let* $\lambda \in P^\vee$, $1 \leq i \leq n$. *Then*

$$Y^\lambda T_i - T_i Y^{s_i \lambda} = \frac{(t^{1/2} - t^{-1/2})(Y^\lambda - Y^{s_i \lambda})}{(1 - Y^{-\alpha_i^\vee})}.$$

PROOF. If this is true for λ and μ, a simple calculation shows that it is true for $\lambda + \mu$ and for $-\lambda$. Hence we may assume that λ is a fundamental weight λ_j. If $j \neq i$, then (3.2) reduces to (2.5), and if $j = i$ it follows from (2.6) and (3.1). $\square$

REMARK. Since $s_i\lambda = \lambda - p\alpha_i^\vee$, where $p = \langle \lambda, \alpha_i \rangle \in \mathbb{Z}$, it follows that the right-hand side of (3.2) is a polynomial in the Y's.

From (3.2) it follows (cf. [**L**]) that

(3.3) *The elements* $T(w)Y^\lambda$ *(resp. the elements* $Y^\lambda T(w)$*), where* $w \in W_0$ *and* $\lambda \in P^\vee$*, form a* K*-basis of* H*.*

Let $A^\vee = K[P^\vee]$ be the group algebra of $P^\vee$ over K. As in Chapter II, the formal exponentials e^λ, $\lambda \in P^\vee$, form a K-basis of $A^\vee$. For each $f \in A^\vee$, say $f = \sum f_\lambda e^\lambda$, let

$$f(Y) = \sum f_\lambda Y^\lambda \in H$$

and let $A^\vee(Y)$ denote the subalgebra of H generated by the Y^λ, $\lambda \in P^\vee$. From (3.3) we have $A^\vee(Y) \cong A^\vee$ and

$$H \cong A^\vee \otimes_K H_0, \tag{3.4}$$

where H_0 is the Hecke algebra of the finite Weyl group W_0, generated by $T_1, \dots, T_n$ subject to the braid relations (2.2) not involving T_0, and the Hecke relations (3.1) (with $i \neq 0$).

If $a = \alpha + nc \in S$, we define

$$e^a = e^{\alpha + nc} = q^n e^\alpha$$

(i.e., we define e^c to be q). The following proposition is due to Cherednik [**C2**]:

(3.5) *The Hecke algebra* H *acts on* $A = K[P]$ *as follows*:

$$T_i e^\mu = t^{1/2} e^{s_i\mu} + (t^{1/2} - t^{-1/2})(1 - e^{a_i})^{-1}(e^\mu - e^{s_i\mu}),$$
$$\omega e^\mu = e^{\omega\mu},$$

where $0 \leq i \leq n$ *and* $\omega \in \Omega$*. Moreover, this representation is faithful.*

We shall sketch a proof. If M is any H_0-module, we can form the induced H_0-module:

$$\mathrm{ind}_{H_0}^H(M) = H \otimes_{H_0} M \cong A^\vee \otimes_K M \tag{3.6}$$

by (3.4). Suppose in particular that M is 1-dimensional and that $T_i x = t^{1/2} x$ for $x \in M$, $1 \leq i \leq n$. From (3.6) the induced module may be identified with $A^\vee$, and by (3.2) the action of T_i $(1 \leq i \leq n)$ on $A^\vee$ is given by

$$T_i e^\lambda = t^{1/2} e^{s_i\lambda} + (t^{1/2} - t^{-1/2})(1 - e^{-\alpha_i^\vee})^{-1}(e^\lambda - e^{s_i\lambda}) \tag{3.7}$$

where $\lambda \in P^\vee$. The operators T_i defined by (3.7) therefore define a representation of H_0 on $A^\vee$, and it is not difficult to show that this representation is faithful. Now H_0 depends only on t and the Weyl group W_0, not on the root system $R^\vee$. We may therefore replace $R^\vee$ by R and $A^\vee$ by A, and the basis $\alpha_1^\vee, \dots, \alpha_n^\vee$ of $R^\vee$ by the opposite basis $-\alpha_1, \dots, -\alpha_n$ of R. This gives us the operators T_i of (3.5) for $1 \leq i \leq n$, and they will by our construction automatically satisfy the braid relations (2.2) not involving T_0, and the Hecke relations (3.1) (for $1 \leq i \leq n$). But then, if we define T_0 as in (3.5), all the relations (2.2), (2.3), and (4.1) will be satisfied, and we have a representation of H on A. □

From (3.5), we have an action of $A^\vee = K[P^\vee]$ on $A = K[P]$, with e^λ $(\lambda \in P^\vee)$ acting as Y^λ. Except in some simple cases it appears not to be possible to make this action explicit, i.e., we cannot calculate $Y^\lambda e^\mu$ explicitly. However, it is possible to calculate the "leading term" of $Y^\lambda e^\mu$, in a sense now to be described, and it will appear that this is sufficient for our purposes.

For this purpose we shall define a partial ordering on the weight lattice P which extends the partial ordering on P^+ defined in Chapter II, §3. If $\lambda \in P$, let λ^+ denote the unique dominant weight in the W_0-orbit of λ, and define (for $\lambda, \mu \in P$)

(3.8) $\lambda \geq \mu$ *if and only if either* (i) $\lambda^+ > \mu^+$, or (ii) $\lambda^+ = \mu^+$ *and* $\mu - \lambda \in Q^+$.

Thus, in a given W_0-orbit, the *antidominant* weight is highest.

Next, for $\mu \in P$, let

$$\rho(\mu) = \frac{1}{2} \sum_{\alpha \in R^+} \varepsilon(\langle \mu, \alpha^\vee \rangle)\alpha \tag{3.9}$$

where $\varepsilon(x) = 1$ if $x > 0$ and $\varepsilon(x) = -1$ if $x \leq 0$; and let

$$\mu^* = \mu + k\rho(\mu). \tag{3.10}$$

Then we have

$$Y^\lambda e^\mu = q^{-\langle \lambda, \mu^* \rangle} e^\mu + \text{ lower terms,} \tag{3.11}$$

for all $\lambda \in P^\vee$ and $\mu \in P$, where by "lower terms" is meant a linear combination of the exponentials e^ν such that $\nu < \mu$.

4. Cherednik's scalar product

The symmetric scalar product $\langle f, g \rangle$ on A defined in Chapter II, §4, is not suitable in the present context, and needs to be modified as follows. Let

$$\Delta' = \prod_{a \in S^+} \frac{1 - e^a}{1 - te^a}. \tag{4.1}$$

Since $t = q^k$ with k a nonnegative integer, it follows from (1.6) that

$$\Delta' = \prod_{\alpha \in R} \prod_{r=0}^{k-1} (1 - q^{\chi(\alpha) + r} e^\alpha) \tag{4.1'}$$

so that

$$\Delta' = \Delta \prod_{\alpha \in R^+} \frac{1 - te^{-\alpha}}{1 - e^{-\alpha}} \tag{4.2}$$

where Δ is defined by Chapter II, (4.1).

Next, if $f \in A$, say $f = \sum f_\lambda e^\lambda$ with coefficients $f_\lambda \in K$, let

$$f^* = \sum f_\lambda^* e^{-\lambda}$$

where f_λ^* is the image of f_λ under the automorphism $q \mapsto q^{-1}$ of K, so that for example $(e^a)^* = e^{-a}$ for $a \in S$.

We now define

$$(f, g) = [fg^* \Delta']_1 \tag{4.3}$$

for $f, g \in A$, where as in Chapter II, §4, the square brackets denote the constant term. This scalar product is nondegenerate and is (almost) hermitian, relative to the involution $q \mapsto q^{-1}$ of K, since $(\Delta')^* = q^{-Nk^2} \Delta'$, where $N = \text{card}(R^+)$.

The advantage of this scalar product is contained in the following proposition [**C2**]:

(4.4) *For each* $w \in W$, *the adjoint of* $T(w)$ *for the scalar product* (4.3) *is* $T(w)^{-1}$, *i.e., we have*

$$(T(w)f, g) = (f, T(w)^{-1}g)$$

for all $f, g \in A$. *In particular, the adjoint of* Y^λ $(\lambda \in P^\vee)$ *is* $Y^{-\lambda}$, *and the adjoint of* $u(Y)$, *where* $u \in A^\vee$, is $u^*(Y)$.

PROOF. It is enough to show that the adjoint of T_i (resp. $\omega \in \Omega$) is T_i^{-1} (resp. ω^{-1}), and this is verified directly from the definitions. □

Finally, when restricted to $A_0 = A^{W_0}$, the scalar product (4.3) is closely related to the symmetric scalar product of Chapter II, §4.

If $f \in A$, say $f = \sum f_\lambda e^\lambda$, let

$$\tilde{f} = \overline{f}^* = \sum f_\lambda^* e^\lambda,$$

so that $\tilde{f}$ is obtained from f by replacing q by q^{-1} (and hence also $t = q^k$ by t^{-1}). Then we have

(4.5) *For all* $f, g \in A_0$,

$$(f, g) = W_0(t)\langle f, \tilde{g}\rangle,$$

where

$$W_0(t) = \sum_{w \in W_0} t^{l(w)}.$$

This follows from (4.2) and the identity [**M2**]

$$\sum_{w \in W_0} \prod_{\alpha \in R^+} \frac{1 - te^{-w\alpha}}{1 - e^{-w\alpha}} = W_0(t). \tag{4.6}$$

5. Another proof of the existence theorem

In this section we shall give an independent (and much simpler) proof of the existence of the orthogonal polynomials P_λ (Chapter II, (5.1)). They will appear as the simultaneous eigenfunctions of the operators $f(Y)$ on A_0, where $f \in A_0^\vee = (A^\vee)^{W_0}$. (It follows from (3.2) that each such operator $f(Y)$ commutes with $T_1, \ldots, T_n$, and hence from (3.5) that $f(Y)$ maps A_0 into A_0.)

If $f \in A^\vee$, say $f = \sum f_\lambda e^\lambda$, and $x \in V$, we define

$$f(x) = \sum f_\lambda q^{\langle \lambda, x\rangle} \tag{5.1}$$

thereby regarding f as a K-valued function on V.

(5.2) *There is a unique* K*-basis* $(P_\lambda)_{\lambda \in P^+}$ *of* A_0 *satisfying the two conditions*

(i) $P_\lambda = m_\lambda +$ *lower terms,*

(ii) $(P_\lambda, m_\mu) = 0$ *for all* $\mu \in P^+$ *such that* $\mu < \lambda$.

Moreover $\widetilde{P}_\lambda = P_\lambda$.

PROOF. Let U (resp. U') be the K-vector space spanned by the m_μ such that $\mu \in P^+$ and $\mu \leq \lambda$ (resp. $\mu < \lambda$). Then U' is a hyperplane in U, and the scalar product remains nondegenerate on restriction to U and to U'. Hence the orthogonal complement of U' in U is 1-dimensional, and contains a unique element P_λ satisfying (i) and (ii).

Next observe that by (4.5) the condition (ii) may be replaced by (ii)′ $\langle P_\lambda, m_\mu\rangle = 0$ *for* $\mu \in P^+$ *such that* $\mu < \lambda$.

We have

$$\widetilde{\Delta} = \prod_{\alpha\in R}\prod_{r=0}^{k-1}(1-q^{-r}e^\alpha) = q^{-Nk(k-1)}\Delta$$

where $N = \operatorname{card}(R^+)$. It follows that $\langle P_\lambda, m_\mu\rangle = 0$ if and only if $\langle \widetilde{P}_\lambda, m_\mu\rangle = 0$. Hence $\widetilde{P}_\lambda$ satisfies the same two conditions (i) and (ii), whence $\widetilde{P}_\lambda = P_\lambda$. □

Now let $f \in A_0^\vee$ and consider $f(Y)m_\mu$, where $\mu \in P^+$. Since

$$m_\mu = e^{w_0\mu} + \text{ lower terms}$$

for the ordering (3.8), where w_0 is the longest element of W_0, it follows from (3.11) and (5.1) that

$$f(Y)m_\mu = f(-(w_0\mu)^*)e^{w_0\mu} + \text{ lower terms.}$$

But $f(Y)m_\mu$ is W_0-symmetric, as remarked at the beginning of this section, and $(w_0\mu)^* = w_0\mu - k\rho = w_0(\mu + k\rho)$ by (3.9) and (3.10). Hence we have

(5.3) $$f(Y)m_\mu = f(-\mu - k\rho)m_\mu + \text{ lower terms.}$$

By (4.3) the adjoint of $f(Y)$ is $f^*(Y)$. Hence if $\mu \in P^+$ and $\mu < \lambda$ we have

$$(f(Y)P_\lambda, m_\mu) = (P_\lambda, f^*(Y)m_\mu)$$

which is zero by (5.3) and (5.2). It follows that $f(Y)P_\lambda$ must be a scalar multiple of P_λ, namely (by (5.3) again)

(5.4) $$f(Y)P_\lambda = f(-\lambda - k\rho)P_\lambda$$

for all $f \in A_0^\vee$. Thus the P_λ diagonalize the action of $A_0^\vee(Y)$ on A_0. Moreover, they are pairwise orthogonal. For we have

$$\begin{aligned} f(-\lambda - k\rho)(P_\lambda, P_\mu) &= (f(Y)P_\lambda, P_\mu) \\ &= (P_\lambda, f^*(Y)P_\mu) \\ &= (P_\lambda, f^*(-\mu - k\rho)P_\mu) \\ &= f(-\mu - k\rho)(P_\lambda, P_\mu) \end{aligned}$$

by (5.4) and (4.3). If $\lambda \neq \mu$ we can choose $f \in A_0$ such that $f(-\lambda - k\rho) \neq f(-\mu - k\rho)$. Hence $(P_\lambda, P_\mu) = 0$ and therefore also $\langle P_\lambda, P_\mu\rangle = 0$ by (4.5), since $\widetilde{P}_\mu = P_\mu$ by (5.2). This completes the proof of the existence theorem.

Remark. Consider again from the present viewpoint the operators D_π and E_π of Chapter II, §6 and §7. By comparing (5.4) with Chapter II, (6.13) and (7.13), it follows that

$$D_\pi = t^{\langle \pi, \rho\rangle} \sum_{w\in W_0} Y^{-w\pi}$$

for $\pi \in P^\vee$ minuscule, and

$$E_\pi = t^{\langle \pi, \rho\rangle} \sum_{w\in W_0} (Y^{-w\pi} - t^{\langle w\pi, \rho\rangle})$$

for $\pi = \varphi^\vee$, φ the highest root of R.

6. The nonsymmetric polynomials E_λ

(6.1) *There is a unique K-basis $(E_\lambda)_{\lambda\in P}$ of A satisfying the two conditions*
(i) $E_\lambda = e^\lambda +$ *lower terms,*
(ii) $(E_\lambda, e^\mu) = 0$ *for all $\mu \in P$ such that $\mu < \lambda$.*

The proof is the same as that of (5.2).

If $f \in A^\vee$, it follows from (4.3) that

$$(f(Y)E_\lambda, e^\mu) = (E_\lambda, f^*(Y)e^\mu)$$

which is zero if $\mu < \lambda$, by (3.11) and (6.1). It follows that $f(Y)E_\lambda$ is a scalar multiple of E_λ, namely (by (3.11) again)

$$f(Y)E_\lambda = f(-\lambda^*)E_\lambda. \tag{6.2}$$

Thus the E_λ diagonalize the action of $A^\vee$ on A, and the same argument as in §5 shows that they are pairwise orthogonal:

$$(E_\lambda, E_\mu) = 0 \quad \text{if } \lambda \neq \mu. \tag{6.3}$$

One shows next that if $\lambda \in P$ is such that $\lambda \neq s_i\lambda$, then T_iE_λ is a linear combination of E_λ and $E_{s_i\lambda}$, with coefficients that can be explicitly computed. From this fact and (6.2), it follows that for each $\lambda \in P^+$ the K-subspace $A(\lambda)$ spanned by the E_μ such that $\mu \in W_0\lambda$ is stable under the action of H (and indeed is an irreducible H-module).

Consider now the operators

$$U^+ = \sum_{w\in W_0} t^{l(w)/2}T(w),$$
$$U^- = \sum_{w\in W_0} (-t)^{-l(w)/2}T(w)$$

on A, where $l(w)$ is the length of $w \in W_0$. We have

$$(T_i - t^{1/2})U^+ = U^+(T_i - t^{1/2}) = 0 \tag{6.4}$$

and

$$(T_i + t^{-1/2})U^- = U^-(T_i + t^{-1/2}) = 0 \tag{6.5}$$

for $1 \leq i \leq n$. From (6.4) it follows that U^+f is W_0-symmetric for all $f \in A$ (but U^-f is *not* W_0-skew, unless $q = 1$). In particular, if $\lambda \in P^+$ then U^+E_λ is a scalar multiple of P_λ (because it has the same defining properties). Hence $P_\lambda \in A(\lambda)$, say

$$P_\lambda = \sum_{w\in W_0\lambda} a_\mu E_\mu$$

and the coefficients a_μ can be calculated explicitly.

Next define, again for $\lambda \in P^+$,

$$Q_\lambda = U^-E_\lambda.$$

If λ is not regular (i.e., if $\langle\lambda, \alpha_i^\vee\rangle = 0$ for some i) then $Q_\lambda = 0$. We have $Q_\lambda \in A(\lambda)$, say

$$Q_\lambda = \sum_{\mu\in W_0\lambda} b_\mu E_\mu$$

and again the coefficients b_μ can be calculated explicitly. In this way both (P_λ, P_λ) and (Q_λ, Q_λ) can be expressed in terms of (E_λ, E_λ) and we obtain (λ dominant regular)

$$\frac{(Q_\lambda, Q_\lambda)}{(P_\lambda, P_\lambda)} = t^{-N} \prod_{\alpha\in R^+} \frac{1-q^{\langle\lambda+k\rho,\alpha^\vee\rangle+k}}{1-q^{\langle\lambda+k\rho,\alpha^\vee\rangle-k}}. \tag{6.6}$$

7. Calculation of $\langle P_\lambda, P_\lambda\rangle$

In this section we shall sketch a proof of the formula (8.3′) of Chapter II, giving the value of the scalar product $\langle P_\lambda, P_\lambda\rangle$. The proof is by induction on k, the case $k = 0$ (or $k = 1$) being trivial. From now on we shall write $P_{\lambda,k}$ and $Q_{\lambda,k}$ in place of P_λ and Q_λ, to stress the dependence on the parameter k, and likewise for the scalar products: $(f, g)_k$ and $\langle f, g\rangle_k$ in place of (f, g) and $\langle f, g\rangle$.

An element $f \in A$ will be called *H_0-symmetric* (resp. *H_0-skew*) if $T_i f = t^{1/2} f$ (resp. $-t^{-1/2} f$) for $1 \le i \le n$. H_0-symmetric is the same as W_0-symmetric, as is clear from (3.5); but H_0-skew is *not* the same as W_0-skew (unless $q = 1$). The polynomials Q_λ introduced in §6 are H_0-skew, and so is the product

$$\pi_k = \prod_{\alpha\in R^+} (e^{\alpha/2} - t^{-1} e^{-\alpha/2}).$$

Multiplication by π_k converts H_0-symmetric into H_0-skew, and indeed the space of H_0-skew elements of A is precisely $\pi_k A_0$. This leads to the following formula connecting the P's and the Q's:

(7.1) *For all $\lambda \in P^+$ we have*

$$P_{\lambda,k+1} = \pi_k^{-1} Q_{\lambda+\rho,k}.$$

In particular, taking $\lambda = 0$, it follows that $Q_{\rho,k} = \pi_k$, so that (7.1) may be written in the alternative form

$$P_{\lambda,k+1} = Q_{\lambda+\rho,k}/Q_{\rho,k}. \tag{7.1′}$$

This may be regarded as a generalization of Weyl's character formula (Chapter II, (3.5)), which is the case $k = 0$.

Next, we have

$$\pi_k^* = \prod_{\alpha\in R^+} (e^{-\alpha/2} - t e^{\alpha/2}) = e^{-\rho} \prod_{\alpha\in R^+} (1 - q^k e^\alpha)$$

from which it follows that

$$\pi_k \pi_k^* \Delta_k' = \Delta_{k+1} \prod_{\alpha\in R^+} \frac{1-t^{-1}e^{-\alpha}}{1-e^{-\alpha}}$$

and hence, by use of the identity (4.6) (with t replaced by t^{-1}), that

$$(\pi_k f, \pi_k g)_k = W_0(t^{-1}) \langle f, \tilde{g}\rangle_{k+1} \tag{7.2}$$

for all $f, g \in A_0$.

From (7.1) and (7.2), since $\widetilde{P}_\lambda = P_\lambda$ and $W_0(t^{-1}) = t^{-N} W_0(t)$, we obtain

$$(Q_{\lambda+\rho,k}, Q_{\lambda+\rho,k})_k = t^{-N} W_0(t) \langle P_{\lambda,k+1}, P_{\lambda,k+1}\rangle_{k+1}. \tag{7.3}$$

If we now put together (7.3) and (6.6) and make use of (4.5), we shall obtain

$$\frac{\langle P_{\lambda,k+1}, P_{\lambda,k+1}\rangle_{k+1}}{\langle P_{\lambda+\rho,k}, P_{\lambda+\rho,k}\rangle_k} = \prod_{\alpha\in R^+} \frac{1-q^{\langle\lambda+(k+1)\rho,\alpha^\vee\rangle+k}}{1-q^{\langle\lambda+(k+1)\rho,\alpha^\vee\rangle-k}}$$

and hence, on replacing k by $k-1$,

$$\langle P_{\lambda,k}, P_{\lambda,k}\rangle_k = \prod_{\alpha\in R^+}\prod_{i=1}^{k-1} \frac{1-q^{\langle\lambda+k\rho,\alpha^\vee\rangle+i}}{1-q^{\langle\lambda+k\rho,\alpha^\vee\rangle-i}}, \tag{7.4}$$

as stated in Chapter II, §12.

The same techniques can be used to calculate the scalar product (E_μ, E_μ) for any $\mu \in P$. To state the result in a convenient form, we introduce the following notation. For each $w \in W_0$ and $x \in V$ let

$$c_w(x) = \prod_{\alpha\in R^+} \frac{\Gamma_q(\langle x,\alpha^\vee\rangle + \chi(w\alpha))}{\Gamma_q(\langle x,\alpha^\vee\rangle + \chi(w\alpha)+k)} \tag{7.5}$$

and

$$c'_w(x) = \prod_{\alpha\in R^+} \frac{\Gamma_q(\langle x,\alpha^\vee\rangle + \chi(w\alpha)-k)}{\Gamma_q(\langle x,\alpha^\vee\rangle + \chi(w\alpha))} \tag{7.5$'$}$$

where $\chi(w\alpha)$ is 0 or 1 according as $w\alpha$ is positive or negative.

(7.6) *Let $\lambda \in P^+$, let $\mu \in W_0\lambda$ and let w be the longest element of W_0 for which $\mu = w\lambda$. Then we have*

$$(E_\mu, E_\mu) = \frac{c'_w(\lambda+k\rho)}{c_w(\lambda+k\rho)}.$$

8. The double affine Hecke algebra and duality

The affine Hecke algebra H is generated by $T_1, \dots, T_n$ and the Y^λ $(\lambda \in P^\vee)$ and acts faithfully on $A = K[P]$, as we have described in §3. For each $\mu \in P$ let $X^\mu : A \to A$ denote the operator of multiplication by e^μ. More generally, if $f \in A$, say $f = \sum f_\mu e^\mu$, let $f(X) = \sum f_\mu X^\mu$, so that $f(X)g = fg$ for $g \in A$.

The *double affine Hecke algebra* $\mathfrak{H}$ is the algebra of K-linear operators on A generated by H and the X^μ, $\mu \in P$.

(8.1) *Let $\mu \in P$, $0 \le i \le n$. Then*

$$X^\mu T_i - T_i X^{s_i\mu} = (t^{1/2} - t^{-1/2})(1 - X^{a_i})^{-1}(X^\mu - X^{s_i\mu}).$$

This follows from (3.5) by a straightforward calculation.

The formula (8.1) is the counterpart of (3.2), with the Y's replaced by X's and the roles of R and $R^\vee$ interchanged. However, there is one significant difference: (8.1), unlike (3.2), holds for T_0 as well as $T_1, \dots, T_n$.

By using (8.1) it can be shown that

(8.2) *The elements $X^\mu T(w) Y^\lambda$ $(\mu \in P, \lambda \in P^\vee, w \in W_0)$ form a K-basis of $\mathfrak{H}$.*

We have defined $\mathfrak{H}$ concretely as an algebra of linear operators on A. Dually, by interchanging R and $R^\vee$, we may define $\mathfrak{H}^\vee \subset \mathrm{End}_K(A^\vee)$. Then we have

(8.3) *The K-linear mapping $\theta : \mathfrak{H} \to \mathfrak{H}^\vee$ defined by*

$$\theta(X^\mu T(w) Y^\lambda) = X^{-\lambda} T(w^{-1}) Y^{-\mu}$$

for $\lambda \in P^\vee$, $\mu \in P$ *and* $w \in W_0$, *is an anti-isomorphism of* K*-algebras (i.e., we have* $\theta(uv) = \theta(v)\theta(u)$ for $u, v \in \mathfrak{H}$).

Thus θ maps Y^λ to $X^{-\lambda}$, X^μ to $Y^{-\mu}$, and T_i to T_i $(1 \le i \le n)$. But θ does *not* map T_0 to T_0. (Indeed, if $\varphi \in R^+$ is the highest root, so that $a_0 = -\varphi + c$, we have $s_0 s_\varphi = \tau(\varphi^\vee)$ and therefore $T_0 T(s_\varphi) = Y^{\varphi^\vee}$, giving $T_0 = Y^{\varphi^\vee} T(s_\varphi)^{-1}$, so that $\theta(T_0) = T(s_\varphi)^{-1} X^{-\varphi^\vee}$.)

The duality theorem (8.3) is due to Cherednik [**C1**], but as far as I am aware there is as yet no complete proof in the literature, apart from the indications in [**C1**]. A pedestrian approach, which can be carried through, is the following: one can define $\mathfrak{H}$ abstractly via generators and relations, the generators being the X's, the Y's and $T_1, \dots, T_n$. However, the defining relations are not symmetrical as between the X's and Y's, and some work is needed to show that they imply their images under θ (which are relations for $\mathfrak{H}^\vee$). An alternative approach is suggested at the end of §9.

Next, we define a K-linear mapping $\eta : \mathfrak{H} \to K$ as follows: if $u \in \mathfrak{H}$ (acting on A) then $u(1) = u(e^0) \in A$ and we define

$$\eta(u) = u(1)(-k\rho^\vee). \tag{8.4}$$

Dually, we have $\eta^\vee : \mathfrak{H}^\vee \to K$ defined by

$$\eta^\vee(v) = v(1)(-k\rho) \tag{8.4'}$$

for $v \in \mathfrak{H}^\vee$.

These two maps are related by

$$\eta = \eta^\vee \circ \theta. \tag{8.5}$$

PROOF. If h is the basis element $X^\mu T(w) Y^\lambda$ as in (8.2), we have $Y^\lambda(1) = Y^\lambda(E_0) = q^{\langle \lambda, k\rho \rangle} \cdot 1$ by (6.2) (since $0^* = -k\rho$); also $T(w)1 = t^{l(w)}1$, so that $h(1) = q^{\langle \lambda, k\rho \rangle} t^{l(w)} e^\mu$, and therefore

$$\eta(X^\mu T(w) Y^\lambda) = q^{\langle \lambda, k\rho \rangle - \langle \mu, k\rho^\vee \rangle} t^{l(w)}. \tag{1}$$

Likewise

$$\eta^\vee(X^{-\lambda} T(w^{-1}) Y^{-\mu}) = q^{-\langle \mu, k\rho^\vee \rangle + \langle \lambda, k\rho \rangle} t^{l(w)} \tag{2}$$

since $l(w^{-1}) = l(w)$. Now (8.5) follows from (1) and (2). □

(8.6) *Let* $u \in \mathfrak{H}$, $v \in \mathfrak{H}^\vee$. *Then*

$$\eta^\vee(\theta(u)v) = \eta(\theta^\vee(v)u)$$

where $\theta^\vee = \theta^{-1} : \mathfrak{H}^\vee \to \mathfrak{H}$.

PROOF. We have

$$\theta(u)v = \theta(u)\theta(\theta^\vee(v)) = \theta(\theta^\vee(v)u)$$

since θ is an anti-isomorphism. Hence (8.6) follows from (8.5). □

The polynomials E_μ ($\mu \in P$) will now be denoted by E_μ^R, to indicate that they are attached to the root system R. Dually we have polynomials $E_\lambda^{R^\vee} \in A^\vee$, where $\lambda \in P^\vee$. We shall apply (8.6) with $u = E_\mu^R(X)$ and $v = E_\lambda^{R^\vee}(X)$. We have $\theta(u) = \overline{E}_\mu^R(Y)$, so that

$$\begin{aligned}\eta^\vee(\theta(u)v) &= (\overline{E}_\mu^R(Y)E_\lambda^{R^\vee})(-k\rho)\\ &= E_\mu^R(\lambda^*)E_\lambda^{R^\vee}(-k\rho)\end{aligned}$$

by (6.2). Likewise

$$\eta(\theta^\vee(v)u) = E_\lambda^{R^\vee}(\mu^*)E_\mu^R(-k\rho^\vee)$$

and therefore we have [**C4**]

$$E_\mu^R(\lambda^*)E_\lambda^{R^\vee}(-k\rho) = E_\lambda^{R^\vee}(\mu^*)E_\mu^R(-k\rho^\vee) \tag{8.7}$$

for all $\lambda \in P^\vee$ and $\mu \in P$.

Let $\widetilde{E}_\mu^R = E_\mu^R/E_\mu^R(-k\rho^\vee)$ and dually $\widetilde{E}_\lambda^{R^\vee} = E_\lambda^{R^\vee}/E_\lambda^{R^\vee}(-k\rho)$. With this renormalization, (8.7) takes the form

$$\widetilde{E}_\mu^R(\lambda^*) = \widetilde{E}_\lambda^{R^\vee}(\mu^*). \tag{8.7$'$}$$

Again, if we take $u = P_\mu^R(X)$ and $v = P_\lambda^{R^\vee}(X)$, where $\mu \in P^+$ and $\lambda \in (P^\vee)^+$, we shall obtain in the same way

$$P_\mu^R(\lambda + k\rho^\vee)P_\lambda^{R^\vee}(k\rho) = P_\lambda^{R^\vee}(\mu + k\rho)P_\mu^R(k\rho^\vee). \tag{8.8}$$

We mention without proof an explicit formula [**C4**] for $E_\mu^R(-k\rho^\vee)$:

(8.9) *Let $\nu \in P^+$, let $\mu \in W_0\nu$, and let w be the longest element of W_0* such that *$\mu = w\nu$, as in* (7.6). *Then*

$$E_\mu^R(-k\rho^\vee) = q^{-\langle \nu, k\rho^\vee\rangle} c_{w_0}(k\rho)/c_w(\nu + k\rho).$$

9. The Fourier transform

We define a pairing $[\ ,\ ] : A^\vee \times A \to K$ as follows:

$$[f, g] = \eta(\overline{f}(Y)g(X)) = (\overline{f}(Y)g)(-k\rho^\vee) \tag{9.1}$$

where $f \in A^\vee$ and $g \in A$. So if $\mu \in P$ we have

$$\begin{aligned}[f, E_\mu^R] &= (\overline{f}(Y)E_\mu^R)(-k\rho^\vee)\\ &= f(\mu^*)E_\mu^R(-k\rho^\vee)\end{aligned} \tag{9.2}$$

by (6.2). It follows that the pairing (9.1) is nondegenerate: for if $f \in A^\vee$ is such that $[f, g] = 0$ for all $g \in A$, then $f(\mu^*) = 0$ for all $\mu \in P$ (since $E_\mu^R(-k\rho^\vee) \neq 0$ by (8.9)) and hence $f = 0$.

Dually, we define

$$[g, f]^\vee = \eta^\vee(\overline{g}(Y)f(X)) = (\overline{g}(Y)f)(-k\rho) \tag{9.3}$$

for $f \in A^\vee$ and $g \in A$. It now follows from (8.6) that

$$[f, g] = [g, f]^\vee. \tag{9.4}$$

Since by (9.2)

$$[E_\lambda^{R^\vee}, E_\mu^R] = E_\lambda^{R^\vee}(\mu^*)E_\mu^R(-k\rho^\vee)$$

we see that (9.4) is just a restatement of (8.7).

Now let $L : A \to A$ be any linear operator. Since the pairing (9.1) is non-degenerate we may define **[C4]** its *Fourier transform* $\widehat{L} : A^\vee \to A^\vee$ by

(9.5) $$[\widehat{L}f, g] = [f, Lg]$$

where $f \in A^\vee, g \in A$. Clearly we have

(9.6) $$(L_1L_2)^\wedge = \widehat{L}_2\widehat{L}_1$$

for any two operators $L_1, L_2 : A \to A$.

The Fourier transforms of the operators $f(Y), g(X)$ and T_i $(1 \le i \le n)$ are given by

(9.7) (i) $f(Y)^\wedge = \overline{f}(X)$,
(ii) $g(X)^\wedge = \overline{g}(Y)$,
(iii) $\widehat{T}_i = T_i$ $(1 \le i \le n)$.

PROOF. (i) If $L = f(Y)$ and $u \in A^\vee, v \in A$, we have

$$[u, Lv] = (\overline{u}(Y)f(Y)v)(-k\rho^\vee) = [h, v]$$

where $h = u\overline{f} = \overline{f}(X)u$. Hence $\widehat{L} = \overline{f}(X)$.

(ii) If $L = g(X)$ then by (9.4) we have

$$\begin{aligned}[u, Lv] &= [g(X)v, u]^\vee \\ &= [v, \overline{g}(Y)u]^\vee = [\overline{g}(Y)u, v]\end{aligned}$$

by (i) above and (9.4) again.

(iii) It is enough to show that $[e^\lambda, T_iE_\mu] = [T_ie^\lambda, E_\mu]$ for $\lambda \in P^\vee$ and $\mu \in P$. We have $T_ie^\lambda = f$ where

$$f = t^{1/2}e^{s_i\lambda} + (t^{1/2} - t^{-1/2})(e^\lambda - e^{s_i\lambda})/(1 - e^{\alpha_i^\vee})$$

so that

$$\begin{aligned}\overline{f}(Y) &= t^{1/2}Y^{-s_i\lambda} + (t^{1/2} - t^{-1/2})(Y^{-\lambda} - Y^{-s_i\lambda})/(1 - Y^{-\alpha_i^\vee}) \\ &= Y^{-\lambda}T_i - (T_i - t^{1/2})Y^{-s_i\lambda}\end{aligned}$$

by (3.2). Hence

$$\begin{aligned}[T_ie^\lambda, E_\mu] &= (\overline{f}(Y)E_\mu)(-k\rho^\vee) \\ &= [e^\lambda, T_iE_\mu] - ((T_i - t^{1/2})q^{\langle s_i\lambda, \mu^*\rangle}E_\mu)(-k\rho^\vee).\end{aligned}$$

Now for any $g \in A$ we have by (3.5)

$$(T_i - t^{1/2})g = \frac{t^{1/2}e^{\alpha_i} - t^{-1/2}}{e^{\alpha_i} - 1}(g - s_ig)$$

and $e^{\alpha_i}(-k\rho^\vee) = q^{-\langle k\rho^\vee, \alpha_i\rangle} = q^{-k} = t^{-1}$. Hence $((T_i - t^{1/2})g)(-k\rho^\vee) = 0$. □

From (9.7) and (9.6) it follows that if $L = Y^\lambda T(w)X^\mu$, where $\lambda \in P^\vee, \mu \in P$ and $w \in W_0$, then $\widehat{L} = X^{-\mu}T(w^{-1})Y^{-\lambda}$. Hence

(9.8) *For all* $L \in \mathfrak{H}$ *we have* $\widehat{L} = \theta(L) \in \mathfrak{H}^\vee$.

It follows that the duality theorem (8.3) is equivalent to the existence of the Fourier transform, satisfying (9.6) and (9.7). In turn, (9.7) was a consequence of (9.4), or equivalently of the symmetry formula (8.7). Now (8.7) can in fact be proved directly, without recourse to duality, along the lines of Koornwinder's proof of Chapter I, (12.7) (which is reproduced in [**M6**], Chapter VI). This therefore gives an independent proof of (8.3).

10. The general case

As in Chapter II, we have restricted ourselves in this chapter to the affine root systems of the type $S(R)$ (1.1) and a single parameter k, which we have assumed throughout to be a nonnegative integer. The general picture is that one can attach to any irreducible affine root system S, reduced or not, families of orthogonal polynomials P_λ, Q_λ, and E_λ as in the text. These depend (apart from q) on as many parameters k as there are orbits in S under the affine Weyl group W_S, and all the results and formulas of this chapter go over to this more general situation. For an irreducible S, the maximum number of orbits is 5, and is attained by the (nonreduced) affine root systems denoted by $C^\vee C_n$ $(n \geq 2)$ in the tables at the end of [**M1**]. Correspondingly, we have orthogonal polynomials P_λ, Q_λ, and E_λ depending on q and five parameters k_i. These P_λ are the orthogonal polynomials defined by Koornwinder [**K**], which are therefore amenable to the Hecke algebra techniques described here.

References

[C1] I. Cherednik, *Double affine Hecke algebras, Knizhnik-Zamolodchikov equations, and Macdonald's operators*, Internat. Math. Res. Notices (1992), 171–180.

[C2] I. Cherednik, *Double affine Hecke algebras and Macdonald's conjectures*, Ann. Math. **141** (1995), 191–216.

[C3] I. Cherednik, *Macdonald's evaluation conjectures and difference Fourier transform*, Inv. Math. **122** (1995), 119–145.

[C4] I. Cherednik, *Nonsymmetric Macdonald polynomials*, Internat. Math. Res. Notices (1995), 483–515.

[HO] G. J. Heckman and E. M. Opdam, *Root systems and hypergeometric functions* I, II, Comp. Math. **64** (1987), 329–373.

[K] T. H. Koornwinder, *Askey-Wilson polynomials for root systems of type BC*, in Hypergeometric Functions on Domains of Positivity, Jack Polynomials and Applications, edited by D. St. P. Richards, Contemp. Math., vol. 138, Amer. Math. Soc., Providence, RI, 1992, pp. 189–204.

[L] G. Lusztig, *Affine Hecke algebras and their graded version*, J. Amer. Math. Soc. **2** (1989), 599–635.

[M1] I. G. Macdonald, *Affine root systems and Dedekind's η-function*, Inv. Math. **15** (1972), 91–143.

[M2] I. G. Macdonald, *The Poincaré series of a Coxeter group*, Math. Ann. **199** (1972), 161–174.

[M3] I. G. Macdonald, *Some conjectures for root systems*, SIAM J. Math. Anal. **13** (1982), 988–1007.

[M4] I. G. Macdonald, *Orthogonal polynomials associated with root systems*, preprint (1987).

[M5] I. G. Macdonald, *A new class of symmetric functions*, Publ. IRMA, Strasbourg, Actes 20e Séminaire Lotharingien (1988), 131–171.

[M6] I. G. Macdonald, *Symmetric functions and Hall polynomials*, 2nd ed., Oxford University Press, Oxford, 1995.

[M7] I. G. Macdonald, *Affine Hecke algebras and orthogonal polynomials*, Séminaire Bourbaki **797** (1995).

[M8] M. L. Mehta, *Basic sets of invariant polynomials for finite reflection groups*, Comm. Algebra **16** (1988), 1083–1098.

[M9] W. G. Morris, *Constant term identities for finite and affine root systems: Conjectures and theorems*, Ph.D. Thesis, University of Wisconsin, Madison, 1982.

[O1] E. M. Opdam, *Root systems and hypergeometric functions* III, IV, Comp. Math. **67** (1988), 21–49, 191–209.

[O2] E. M. Opdam, *Some applications of hypergeometric shift operators*, Inv. Math. **98** (1989), 1–18.

[O3] E. M. Opdam, *Harmonic analysis for certain representations of graded Hecke algebras*, Acta Math. **175** (1995), 75–121.